T0257440

Sevick's
Transmission Line
Transformers

Sevick's Transmission Line Transformers

Theory and Practice

5th Edition

Raymond A. Mack and Jerry Sevick

SCITECH
PUBLISHING
an imprint of the IET

Edison, NJ scitechpub.com

an imprint of the IET

Published by SciTech Publishing, an imprint of the IET.
www.scitechpub.com
www.theiet.org

Fourth edition 2001
Fifth edition 2014

Editor: Dudley R. Kay

10 9 8 7 6 5 4 3 2 1

ISBN 978-1-89112-197-5 (hardback)
ISBN 978-1-61353-046-7 (PDF)

Typeset in India by MPS Limited
Printed in the US by Integrated Books International
Printed in the UK by CPI Group (UK) Ltd, Croydon

I missed the chance to dedicate my first book to my wife, Debby, so this serves to dedicate both books to her. This work would not have occurred without her endless accommodations to my schedule and her constant encouragement to focus on what is important. I am blessed to have the best wife in the history of the universe.

It is dedicated to Debby's mother, Katherine, who was understanding that we were not able to visit nearly as often as needed. She is an inspiration to her children, their spouses, and her grandchildren.

It is dedicated to my parents, Ray and Helen. I now realize, after putting three kids of my own through college, just how much they had to sacrifice so that I could have an education that enables me to write.

Contents

Preface to the 5th Edition

Jerry Sevick produced early editions of this book as a result of his research into wideband transformers for matching vertical antennas to 50 Ω transmission line. The first edition appeared in 1987. He updated the book several times to improve its content and presentation, and the most recent edition was printed in 2001. Sevick was working on the present 5th edition when he passed away in 2009. He intended to reorganize the book to focus first on theory of operation and then reprise the practical, handbook aspects with specific designs and measurements.

The first four editions read more like a handbook than a textbook. Theory was dispersed throughout, but it was not conducive to easy learning. This revision groups the theory into the first five chapters, and the remainder of the book offers practical designs.

The first three printings were entirely Sevick's work, and the 4th edition added a chapter by Gary Breed on equal delay transformers, which carries over to the present work in Chapter 11. I have added new material, so to distinguish which author is responsible for comments, original work is directly attributed to Sevick and first-person references are to me.

My research into the properties of ferrite materials went beyond Sevick's work. I also investigated modern sources for ferrite materials and copper wire. Unfortunately, a large number of the suppliers that existed when the first edition of this book was published no longer exist, have been purchased by other companies, or otherwise do not supply the parts Sevick used. The sections on material suppliers and the test equipment chapter have been updated to reflect the significant changes since 1987. However, the practical chapters still keep Sevick's initial designs and measurements.

I owe a debt to Sevick's colleagues at AT&T Bell Laboratories, M. D. Fagan and C. L. Ruthroff, who assisted Sevick considerably with the chapters on practical transformer construction and performance.

I wish to thank Elna Magnetics and Fair-Rite Products Corporation for their assistance in providing updated ferrite materials. Fair-Rite was helpful in pointing me to the work of Jim Brown, who then led me to E. C. Snelling's research.

I also wish to thank Kristi Bennett, who used her considerable talent to adjust my words so they make sense and read well. Editors never get sufficient credit for their contributions to books.

Raymond A. Mack
Round Rock, Texas

In Memoriam

Jerry Sevick, W2FMI—renowned for his research and publications related to short vertical antennas and transmission line transformers—passed away on November 29, 2009, at the age of 90.

Jerry was a graduate of Wayne State University and a member of its Athletic Hall of Fame. During World War II, he served as a pilot in the US Army Air Corps. In 1952, he graduated from Harvard University with a doctorate in Applied Physics with a dissertation titled "An Experimental and Theoretical Investigation of Back-Scattering Cross Sections." From 1952–1956, he returned to Wayne State University to teach physics and also became a local weather forecaster on WXYZ-TV7. In 1956, he joined AT&T Bell Laboratories where he supervised groups working in high-frequency transistor and integrated-circuit development, reliability, applications engineering and high-speed PCM; later, he served as Director of Technical Relations at the company and retired in 1985.

During 1971–1981 Jerry authored 10 *QST* articles on antenna-related topics with the majority covering vertical antennas, especially shortened verticals. However, he also covered radial systems and ground conductivity, broadband matching networks and impedance bridges. In the course of designing networks to match coaxial cable to short ground mounted vertical antennas, Jerry looked at the transmission line transformer as a possible vehicle. He undertook the characterization and design of transformers for low impedance applications, resulting in this book, originally published in 1987 by the ARRL. Jerry's research is also reflected in his publications *Understanding, Building and Using Baluns and Ununs* and *The Short Vertical Antenna and Ground Radial*.

Accolades include the ARRL Hudson Division Technical Achievement Award, which he received in 2004, while serving as an ARRL Technical Advisor, and the Dayton Hamvention Technical Excellence Award, received in 2005, respectively. The Hamvention Awards Committee noted that Sevick's April 1978 QST article on short ground-radial systems "now serves as the world's standard for earth conductivity measurements."

"Jerry embodied the old-fashioned amateur spirit of innovation by experiment, applying his many years of experience as a Bell Labs researcher to a retirement project analyzing the performance of short vertical antennas," said Gary Breed, K9AY, Jerry's collaborator and editor. "That work led him to the study of transmission line transformers for which he became well known in both the ham and professional radio engineering communities. He brought a little-known piece of technology to the forefront and worried until the end whether enough people understood the principles behind the operation of these devices."

Dudley Kay, our present editor, observed: "I don't know anybody who's ever had an unkind word for Jerry Sevick. Quite to the contrary, I have only heard admiration and awe for his kind and warm-hearted nature and his enthusiastic willingness to answer any question and share his knowledge, findings, and even his own lingering questions in as much detail as a person was willing to hear! Among communications engineers he is considered among the giants in the field, and many owe him a debt of gratitude." I heartily concur.

Raymond A. Mack

Chapter 1

Transformer Basics

1.1 Introduction

There are two basic methods for constructing broadband impedance matching transformers. One employs the conventional magnetically coupled transformer that transmits energy to the output circuit by flux linkage; the other uses a transmission line transformer to transmit energy by transverse transmission line mode. Conventional transformers have been constructed to perform over wide bandwidths by exploiting high magnetic efficiency of modern materials. Losses on the order of 1 dB can exist over a range from a few kilohertz to over 200 MHz. Throughout a considerable portion of this range, the losses are only 0.2 dB. Transmission line transformers exhibit far wider bandwidths and much greater efficiencies. The stray inductances and interwinding capacitances are generally absorbed into the characteristic impedance of the transmission line. The flux is effectively canceled out in the core with a transmission line transformer, so extremely high efficiencies are possible over large portions of the passband—losses of only 0.02–0.04 dB with certain core materials.

A full model of a conventional transformer is presented in Figure 1-1. Multiple parasitic elements are affecting both low and high frequency operation. Low frequency operation is controlled by the magnetizing inductance (L_M) in parallel with the ideal transformer. As frequency decreases, the current flows mostly through the low impedance inductance (L_M) rather than the ideal transformer. High frequency operation is governed by the capacitances (C_P, C_S, and C_{PS}), leakage inductances (L_P and L_S), and core losses (R_C). As frequency increases, the output voltage and current become out of phase and the losses of the core increase. R_P and R_S are the copper losses of the respective windings. They increase with increasing frequency due to skin effect. R_P and R_S also increase with increasing temperature, so higher power applications will have higher losses.

Figure 1-2 shows the construction of two conventional transformers on a double "U" core. The windings are physically separated on the core in the first example, so the only linkage from primary to secondary occurs through the shared flux in the core. C_P and C_S in the model occur because each turn of the winding is in proximity to the adjacent turns. A very small capacitance exists between each pair of turns, and each capacitor is in series with the next one around the winding. The capacitance can be quite small, but a capacitor of only 10 pF has an impedance

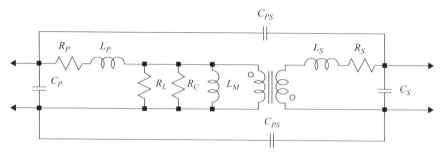

Figure 1-1   The schematic shows a complete model of a magnetically coupled transformer.

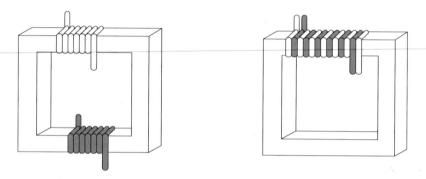

Figure 1-2   Two different methods of magnetically coupled transformer construction: (Left) This transformer minimizes capacitive coupling with all coupling by magnetic flux. (Right) This transformer has coupling by both capacitance and magnetic flux.

of 159 at 100 MHz. Another method of transformer design in the figure winds the secondary on top of the primary. This construction reduces L_P and L_S but increases C_{PS}. The other advantage of this construction is that flux linkage is improved at higher frequencies where the transformer tends to look more like an air core transformer with an absorber in the middle.

A basic transmission line transformer with an unbalanced input and a balanced load is illustrated in Figure 1-3. Two pieces of equal length transmission line are connected in parallel at the input side and in series on the output side. If a transmission line is terminated in its characteristic impedance, the input side appears to be Z_0 regardless of the length of the transmission line (within limits). In the example in Figure 1-3, Z_0 and each half of the load are 100 Ω. The impedance on the input side is 50 Ω because we have two 100 Ω impedances in parallel.

Oliver Heaviside used Maxwell's equations in the late nineteenth century to develop the mathematical expressions for transmission lines. Those equations show that the load is isolated from the input on a transmission line that is longer than about 0.1 wavelength. At that point, the distributed inductance and distributed

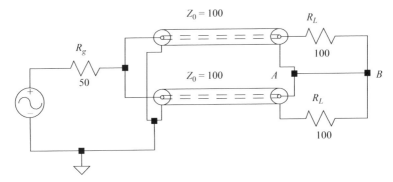

Figure 1-3   *A schematic showing a Guanella 1:4 transformer. The connection between points A and B is used when a load is center tapped. For a 200 Ω load the center tap is omitted.*

capacitance combine to produce the effect we know as *characteristic impedance*. The input energy is transmitted down the line as an electromagnetic field completely contained within the transmission line. For that reason, placing a magnetic core around a transmission line will have no effect on the field inside the line. However, as the length becomes less than 0.1 wavelength, the field is no longer contained within the line so both conductors contribute magnetic flux in a core placed around the line. This external flux converts the line and core combination from a transmission line to a conventional transformer. Thus, the power ratings of transmission line transformers are determined more by the ability of the transmission lines to handle the voltages and currents at high frequencies and by the properties of the core at low frequencies.

The earliest presentation on transmission line transformers was by Gustav Guanella in 1944 [1]. He proposed the concept of coiling transmission lines to form a choke that would reduce the undesired mode in balanced-to-unbalanced (*balun*) matching applications. Before this time, this type of device was constructed from quarter- or half-wavelength transmission lines and, as such, had very narrow bandwidths. By combining coiled transmission lines in parallel-series arrangements, he was able to demonstrate broadband baluns with ratios of $1:n^2$, where n is the number of transmission lines.

Other writers followed with further analyses and applications of the balun transformer [2–8]. In 1959, C. L. Ruthroff published another significant work on this subject [9]. By connecting a single transmission line such that a negative or a positive potential gradient existed along the length of the line, he was able to demonstrate a broadband 1:4 balun, or unbalanced-to-unbalanced (*unun*) transformer. He also introduced the hybrid transformer in his paper. Many extensions and applications of his work were published and are included in the reference list [10–28]. The original Guanella article is reproduced in Appendix A, and the original Ruthroff article is reproduced in Appendix B.

In general, it can be said that the transmission line transformer enjoys the advantage of higher efficiency, greater bandwidth, and simpler construction.

The conventional transformer, however, remains capable of DC isolation. The purpose of this chapter is twofold: to review Guanella's and Ruthroff's approaches and to present additional material to form a basis for the chapters that follow.

Jerry Sevick lamented in the second edition of this book that many readers mistakenly consider the transformer designs in the book to be conventional magnetically coupled transformers. I attribute this to the graphics that depict each conductor of the transmission line as an inductor. I have modified the graphics to show all two-conductor transmission lines as loaded wire lines or coaxial cables. In general, it is equally correct to build a transmission line transformer with coaxial cable as with a parallel line. However, some of the effects that Sevick has observed can be attributed to placing a magnetic material in proximity to a parallel wire transmission line. Until coax came into common use, it was well known that all metal must be kept away by at least four to five times the wire spacing to prevent distorting the signal in the parallel line. The result is that the two-, three-, and four-wire transmission lines more closely resemble coupled microstrip lines than parallel wire lines. Therefore, the core is an integral part of the circuit throughout the useful frequency range. Further, I believe Sevick was mistaken in his understanding that many of the designs presented are strictly transmission line transformers. The Ruthroff designs, in particular, rely on true magnetic transformer action for significant portions of their band of operation. However, his experimental work is still quite relevant!

1.2 The Basic Building Block

The single bifilar winding, shown in Figure 1-4, is the basic building block for understanding and designing transmission line transformers. Higher orders of windings (e.g., trifilar, quadrifilar) also perform in a similar transmission line fashion and will be discussed later.

The circuit in Figure 1-4 can perform four different functions depending on how the output load, R_L, is grounded: (1) a phase inverter when a ground is connected to terminal 4; (2) a balun when the ground is at terminal 5 or left off entirely (a floating load); (3) a simple delay line when a ground is at terminal 2; and (4) a "bootstrap" when $+V_1$ is connected to terminal 2. The operation of these four

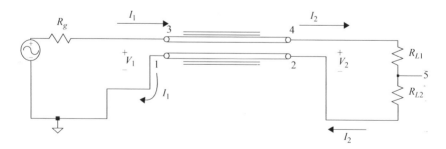

Figure 1-4   The schematic shows the transmission line transformer basic building block.

functions can be explained by simple transmission line theory and the choking reactance of the transmission lines. The latter, which isolates the input from the output, is usually obtained by coiling the transmission line around a ferrite core or by threading the line through ferrite beads. The objectives, in practically all cases, are to have the characteristic impedance (Z_0) of the transmission line equal to the value of the load (R_L), which is called the optimum characteristic impedance, and to have the choking reactance of the transmission line much greater than R_L (and hence Z_0). Meeting these objectives results in a "flat" line and hence maximum high frequency response and maximum efficiency since conventional transformer currents are suppressed. In the final analysis, the maximum high frequency response is determined by the parasitic elements not absorbed into the characteristic impedance of the line, and the efficiency is affected by the properties of the ferrites when used in transmission line transformer applications.

A deeper understanding of transmission line transformers can be gained by noting the longitudinal potential gradients that exist with the following four circuits.

1.2.1   Phase Inverter

By connecting a ground to terminal 4, a negative potential gradient of $-V_1$ is established from terminal 3 to 4. The gradient from terminal 1 to 2 is $-V_2$. For a matched load, $V_1 = V_2$. If the reactance of the windings (or a straight transmission line loaded with beads) is much greater than R_L, then only transmission line currents flow and terminal 2 is at a $-V_2$ potential. When the reactance is insufficient, a shunting, conventional current will also flow from terminal 3 to 4, resulting in a drop in the input impedance and the presence of flux in the core. As the frequency is decreased, the input impedance approaches zero.

1.2.2   Balun

By connecting a ground to terminal 5, a negative potential gradient $-(V_1 - V_2/2)$ is established from terminal 3 to 4 and $-V_2/2$ from terminal 1 to 2. With a matched load, $V_1 = V_2$ and the output is balanced to ground. When the reactance fails to be much greater than R_L, conventional transformer current will flow and eventually, with decreasing frequency, the input impedance approaches $R_L/2$. When the load is "floating," the currents in the two windings are always equal and opposite. At very low frequencies, where the reactance of the windings fails to be much greater than R_L, the isolation of the load is inadequate to prevent conventional transformer current (which could be an antenna current) when the load is elevated in potential. This bifilar balun, which was first proposed by Guanella [1], is completely adequate for most 1:1 balun applications when the reactance of the windings (or beaded straight transmission lines) is much greater than R_L.

1.2.3   Delay Line

By connecting the ground to terminal 2, the potential gradient across the bottom conductor is zero. With a matched load, the gradient across the top conductor is also zero. Under these conditions, the longitudinal reactance of the conductors plays no role. The transmission line simply acts as a delay line and does not require winding about a core or the use of ferrite beads. This delay function plays the most important role in obtaining the highest frequency response in unun transformers.

1.2.4   Bootstrap

Probably the most unlikely circuit schematic with the basic building block is the one where $+V_1$ is also connected to terminal 2 (i.e., terminal 3 is connected to terminal 2). By this type of connection, a positive potential gradient of V_1 is established across the bottom conductor and of $+V_2$ across the top conductor. When the bottom of R_L is connected to ground instead of terminal 2, a voltage of $(V_1 + V_2)$ exists across its terminals. This bootstrap connection, in which the transmission line shares a part of the load, is the way Ruthroff [9] obtained his 1:4 unun transformer.

1.3   Designing a Magnetic Transformer

It is instructive to look at the loss calculations for a magnetic transformer. The following is a design for a 1 to 3 MHz manganese-zinc ferrite (MnZn) transformer but the process is the same for a nickel-zinc ferrite (NiZn) core.

Our example design is a 1:4 step-up transformer for 50:200 Ω conversion with operation at 2000 W peak envelope power (PEP) continuous operation. We choose a 3F5 core because it is characterized for operation at 3 MHz. Transformer design is always an iterative process to ensure that the core and number of windings will produce a transformer that does not melt at the first application of full power. The first step is to choose a core shape and size as a guess. Ferroxcube gives us a starting point in Table 1 of its 2009 *Application Handbook* [29]. Its suggestion is that an E65 core is required for power levels above 500 W. The next guess is for a specific set of windings. Our goal is always to use the minimum number of windings to minimize copper loss and stray inductance. We will start with four turns of no. 16 magnet wire for the primary and eight turns of no. 16 for the secondary. The voltage applied to the primary is set by the power level and impedance. For 2000 W at 50 Ω, we have 316 V RMS. The important parameter is volts/turn. For our transformer, we have 79 V/turn.

The first calculation is to determine the peak flux density of the transformer:

$$B = \frac{E \times 10^2}{4.44 \times N \times A_e \times f} = \frac{316 \times 10^2}{4.44 \times 4 \times 5.4 \times 3} = 110\ G = 11\ mT$$

where

   E = voltage (RMS)
   A_e = effective cross sectional area of the core (cm^2)
   f = frequency in MHz
   N = number of turns

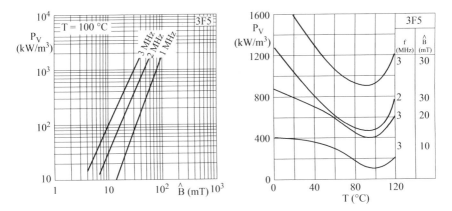

Figure 1-5   *These graphs show the power density data for Ferroxcube 3F5*
material. (Left) Curves show data versus frequency at a constant
temperature. (Right) Curves show power density versus core
temperature, flux density, and frequency.

From the data sheet for the E65/32/27 core, the equivalent area is 540 mm^2, or 5.4 cm^2. Figure 1-5 shows the specific power loss versus flux density and frequency: 11 mT yields approximately 110 mw/cm^3. From the data sheet, the volume of the E65 core is 79,000 mm^3, or 79 cm^3, so the power loss is 8.7 W, or 99.3% efficient. This is a very good transformer. If we had chosen two turns for the primary, the volts/turn would double to 158 V/turn and double the flux density to 22 mT. However, the power loss would be significantly higher at 550 mw/cm^3. Now the power loss would be 43.5 W! This is an object lesson in why the volts/turn value is so important for a transformer.

Let's look at one more variation on our transformer and reduce the maximum power to 200 W. The voltage is now 100 V. The four-turn transformer will consume 1.1 W, and the two-turn transformer will consume only 7 W. We are transmitting only 1/10 the power, but the heat lost in the transformer has only decreased by 1/5.

You may wonder how we can make a transformer with so few turns. How do we make sure the magnetizing inductance is high enough? As frequency decreases, don't we need more turns to maintain enough impedance to keep the transformer operating correctly? All of these concerns are correct, but they are implicitly accounted for by the formula for peak flux density and by the specific power loss curves. Notice that as frequency decreases, peak flux density rises while keeping turns and core area constant. Since frequency is in the denominator, it will have a hyperbolic effect on flux density. Peak flux density does not have a component related to permeability. Also notice that the specific power loss curves increase power loss with decreasing frequency and move the curve to the right on the graph. The movement of the curve to the right in the graph is the implicit contribution of permeability to magnetizing inductance. Both of these effects provide the implicit need to increase the inductance (number of turns and also impedance) as frequency decreases.

The loss calculations described in the previous example are also relevant when we design a transmission line transformer. At low frequencies, a transmission line transformer degenerates into a magnetic transformer, so peak flux density becomes an important factor in transformer operation.

References

[1] Guanella, G., "Novel Matching Systems for High Frequencies," *Brown-Boveri Review*, Vol. 31, Sep. 1944, pp. 327–329.

[2] Fubini, E. G., and P. J. Sutro, "A Wide-Band Transformer from an Unbalanced to a Balanced Line," *Proceedings of the IRE*, Vol. 35, Oct. 1947, pp. 1153–1155.

[3] Rudenberg, H. G., "The Distributed Transformer," Research Division, Raytheon Manufacturing Co., Waltham, MA, Apr. 1952.

[4] Rochelle, R. W., "A Transmission-Line Pulse Inverter," *Review of Scientific Instruments*, Vol. 23, No. 6, 1952, p. 298.

[5] Lewis, I. A. D., Note on "A Transmission Line Pulse Inverter," *Review of Scientific Instruments*, Vol. 23, No. 12, 1952, p. 769.

[6] Brennan, A. T., "A UHF Balun," RCA Laboratories Division, Industry Service Laboratory, LB-911, May 5, 1953.

[7] Talkin, A. I., and J. V. Cuneo, "Wide-Band Transformer," *Review of Scientific Instruments*, Vol. 28, No. 10, 808, Oct. 1957.

[8] Roberts, W. K., "A New Wide-Band Balun," *Proceedings of the IRE*, Vol. 45, Dec. 1957, pp. 1628–1631.

[9] Ruthroff, C. L., "Some Broad-Band Transformers," *Proceedings of the IRE*, Vol. 47, Aug. 1959, pp. 1337–1342.

[10] Turrin, R. H., "Broad-Band Balun Transformers," *QST*, Aug. 1964, pp. 33–35.

[11] Matick, R. E., "Transmission Line Pulse Transformers—Theory and Applications," *Proceedings of the IEEE*, Vol. 56, No. 1, Jan. 1968, pp. 47–62.

[12] Pitzalis, O., and T. P. Couse, "Practical Design Information for Broadband Transmission Line Transformers," *Proceedings of the IEEE*, Apr. 1968, pp. 738–739.

[13] Pitzalis O., and T. P. Couse, "Broadband Transformer Design for RF Power Amplifiers," US Army Technical Report ECOM-2989, Jul. 1968.

[14] Turrin, R. H., "Applications of Broad-Band Balun Transformers," *QST*, Apr. 1969, pp. 42–43.

[15] Pitzalis, O., R. E. Horn, and R. J. Baranello, "Broadband 60-W Linear Amplifiers," *IEEE Journal of Solid State Circuits*, Vol. SC-6, No. 3, Jun. 1971, pp. 93–103.

[16] Krauss, H. L., and C. W. Allen, "Designing Toroidal Transformers to Optimize Wideband Performance," *Electronics*, Aug. 16, 1973.

[17] London, S. E., and S. V. Tomeshevich, "Line Transformers with Fractional Transformation Factor," *Telecommunications and Radio Engineering*, Vols. 28–29, Apr. 1974.

[18] Granberg, H. O., "Broadband Transformers and Power Combining Techniques for RF," Motorola Application Note AN-749, 1975.

[19] Sevick, J., "Simple Broadband Matching Networks," *QST*, Jan. 1976, p. 20.

[20] Sevick, J., "Broadband Matching Transformers Can Handle Many Kilowatts," *Electronics*, Nov. 25, 1976, pp. 123–128.

[21] Blocker, W., "The Behavior of the Wideband Transmission Line Transformer for Nonoptimum Line Impedance," *Proceedings of the IEEE*, Vol. 65, 1978, pp. 518–519.

[22] Sevick, J., "Transmission Line Transformers in Low Impedance Applications," *MIDCON* 78, Dec. 1978.

[23] Dutta Roy, S. C., "Low-Frequency Wide-Band Impedance Matching by Exponential Transmission Line," *Proceedings of the IEEE* (Letter), Vol. 67, Aug. 1979, pp. 1162–1163.

[24] Dutta Roy, S. C., "Optimum Design of an Exponential Line Transformer for Wide-Band Matching at Low Frequencies," *Proceedings of the IEEE* (Letter), Vol. 67, No. 11, Nov. 1979, pp. 1563–1564.

[25] Irish, R. T., "Method of Bandwidth Extension for the Ruthroff Transformer," *Electronic Letters*, Vol. 15, Nov. 22, 1979, pp. 790–791.

[26] Kunieda, H., and M. Onoda, "Equivalent Representation of Multiwire Transmission-Line Transformers and Its Applications to the Design of Hybrid Networks," *IEEE Transactions on Circuits and Systems*, Vol. CAS-27, No. 3, Mar. 1980, pp. 207–213.

[27] Granberg, H. O., "Broadband Transformers," *Electronic Design*, Jul. 19, 1980, pp. 181–187.

[28] Collins, R. E., *Foundations for Microwave Engineering*, New York: McGraw Hill, 1966, Chap. 5.

Chapter 2

Ferrite Materials

2.1 Introduction

With T. Takei's discovery of magnetic ferrites in Japan and the excitement brought about by the 1947 publication of work done at the Philips Research Laboratories (later called Ferroxcube) in the Netherlands during World War II, new and improved devices emerged in the field of magnetics [1,2]. The chemical formula for ferrites is MFe_2O_4, where M stands for any of the divalent ions magnesium, zinc, copper, nickel, iron, cobalt, or manganese or their mixture. Except for compounds containing divalent iron ions, ferrites can be made with bulk resistivities in the range of 10^2 to 10^9 Ω-cm compared with 10^{-5} Ω-cm for the ferromagnetic metals (e.g., powdered iron). This increase in bulk resistivity represents a major step forward for applications in frequency ranges heretofore unobtainable.

Ferrite compositions are made using ceramic technology. This involves intimate mixing fine powders of appropriate oxides, compressing the mixture, and firing it in carefully controlled atmospheres at temperatures of about 1100 to 1200°C. Single crystals have been made by several techniques. By using different combinations of oxides and variations in ceramic processing, the mixtures can be tailored to fit a wide variety of technical requirements. In fact, ferrites with similar specifications by various manufacturers have been found to exhibit different efficiencies in transmission line transformers. This is because the ferrite is not defined completely by its chemistry and crystal structure but also by its processing. These parameters are primarily powder preparation, compact formation, sintering, and machining the ferrite to its final shape.

By the early 1950s, it was generally recognized that inductor cores of Permalloy dust had reached the point of diminishing returns in their application to higher and higher frequencies. In 1952, F. J. Schnettler and A. G. Ganz of Bell Labs contributed significantly to this problem's solution when they developed a high permeability manganese-zinc (MnZn) ferrite for use at telephone carrier frequencies of 100 kHz and higher [3]. This material has also found widespread use at lower frequencies in power transformers, flyback transformers, and deflection yokes.

In the 1960s and 1970s, in response to the rising need for high quality linear devices in the transmission area Bell Labs scientists also made several important advances in linear ferrite properties. The use of cobalt additives and carefully controlled cooling made these advances possible. As a result, a process was developed for making suitable nickel-zinc (NiZn) ferrites capable of operating up to 500 MHz [4]. This form of ferrite is the best one for high power transmission line transformers, and it is commercially available.

The employment of ferrite in inductors and transformers for carrier frequencies has equally impacted communications and microwave and computer technology. The availability of magnetic oxides eventually led to the large family of non-reciprocal magnetic devices that play a key role in microwave technology. The materials effort is credited largely to L. G. Van Uitert of Bell Labs. He proposed the substitution of nonmagnetic ions for magnetic ions in the ferrite structure to reduce internal fields and thereby lower the ferromagnetic resonance frequency.

In the computer field, A. Schonberg of Steatit-Magnesia AG in Germany and workers at MIT's Lincoln Laboratories found a family of magnesium-manganese (MgMn) ferrites with remarkably square hysteresis loops for use in memory and other computer and switching applications. These devices subsequently gave way to the semiconductor logic and memory circuits of the mid-1970s.

Although considerable information is available on the theory and application of transmission line transformers, dating back to the classic papers of Guanella in 1944 and Ruthroff in 1959, virtually no investigations have been made on the use of ferrites in power applications [5,6]. Discussions between Sevick and scientists and engineers from major laboratories working in the ferrite field confirm this lack of development. The following sections will show Sevick's results on readily available ferrites and their use in high power transmission line transformers. It is of some interest to note the differences in the properties of ferrites resulting from variations in processing techniques used by the different manufacturers. We will also look at some recent experiments regarding ferrite properties at high power and high frequency.

2.2   Ferrite Physical Properties

Our primary interest in ferrites is their magnetic operation. However, other physical properties have secondary effects on performance as transmission line transformers. As mentioned already, a ferrite is a ceramic composed of many fundamental elements. A major difference between magnetic materials such as powdered iron and ferrite is resistivity. Powdered iron is a conductor and has very low resistivity on the order of 10^{-3} Ω-m that is minimized in high frequency core applications by surrounding very small iron particles with an insulating coating. The coating reduces eddy current losses by reducing the volume of the magnetic particles. However, the coating forms a distributed air gap that radically reduces effective permeability and makes powdered iron inappropriate for broadband transformers.

Ferrites are semiconductors with resistivity on the order of 10 Ωm for MnZn materials and 10^6 Ωm for NiZn materials. The resistivity within the individual domains is rather small, but the oxides at the crystal boundaries raise the bulk resistivity. The oxides also provide some capacitance between individual crystals.

Table 2-1   Core Properties versus Temperature

Temperature (°C)	MnZn Resistivity (Ω-m)	NiZn Resistivity (Ω-m)
−20	~ 10	$\sim 10^8$
0	~ 7	$\sim 5 \times 10^7$
20	~ 4	$\sim 10^7$
50	~ 2	$\sim 10^6$
100	~ 1	$\sim 10^5$

Table 2-2   Core Properties versus Frequency

Frequency (MHz)	MnZn Resistivity (Ω-m)	NiZn Resistivity (Ω-m)	MnZn Permittivity (ε_r)	NiZn Permittivity (ε_r)
0.1	~ 2	$\sim 10^5$	$\sim 2 \times 10^5$	~ 50
1	~ 0.5	$\sim 5 \times 10^4$	$\sim 10^5$	~ 25
10	~ 0.1	$\sim 10^4$	$\sim 5 \times 10^4$	~ 15
100	~ 0.01	$\sim 10^3$	$\sim 10^4$	~ 12

Resistivity varies with both frequency and temperature. Table 2-1 shows ferrite properties versus temperature, and Table 2-2 lists ferrite properties versus frequency. Loss due to bulk resistivity increases with frequency as the effective capacitive impedance between crystals in the lattice decreases and allows the bulk resistivity to absorb more energy.

The base permittivity of all ferrites is approximately 10. The isolating materials on the grain boundaries also have permittivity approximately 10. However, the effective bulk permittivity varies with frequency because of the conductivity of the crystals. In essence, the ferrite is a complicated network of lossy capacitors. This property is important in transmission line transformers because it combines with the permeability to affect the characteristic impedance of the transmission line.

2.3   Ferrite Permeability

A ferrite is a ceramic composed of small crystals that typically have diameters of 10 to 20 μm. The crystals contain magnetic domains in which the molecular magnets are aligned, but the magnetic domains are not aligned. When an external field is applied to the ferrite, the magnetic domains begin to align with the field. Energy from the applied field is absorbed by the magnetic domains before the domains begin to align. This delay in alignment accounts for the shape of the hysteresis loop, which plots applied field versus magnetic flux in the ferrite. Likewise, as the field is reduced, the alignment of the domains tends to remain. When the applied field reaches zero, the domains will maintain a small amount of residual magnetism.

Permeability (μ) is the ratio of flux density to applied field. When the applied field varies with time, not all of the energy is returned to the circuit but is lost as heat.

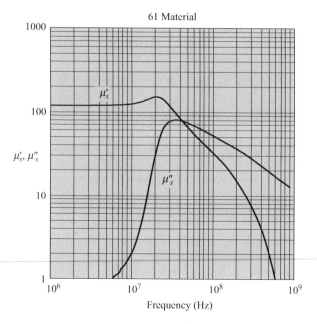

Figure 2-1   These curves show the complex permeability data for Fair-rite material 61. μ'_s corresponds to the inductive component of permeability. μ''_s corresponds to the real or resistive loss portion of permeability.

An inductor will therefore have an effective impedance that consists of an ideal inductance and a resistance. Permeability of a core is expressed as a complex entity that accounts for the inductive component as well as the loss component:

$$\mu = \mu'_s - j\mu''_s$$

Figure 2-1 shows the complex permeability for material 61. The inductance without a core (L_0) is increased by the amount of the permeability of the core:

$$L = L_0 \left(\mu'_s - j\mu''_s \right)$$

And impedance becomes

$$j\omega L_0 \mu'_s - j^2 \omega L_0 \mu''_s = j\omega L_0 \mu'_s + \omega L_0 \mu''_s$$

The loss impedance becomes a positive real value because j is squared. The phase shift caused by the magnetic losses is expressed by

$$\tan \delta_m = \frac{R_s}{\omega L_s} = \frac{\mu''_s}{\mu'_s}$$

Ordinarily, magnetic material data sheets show the impedance of an inductor (Figure 2-2), where the inductance is due to μ'_s and the resistance is due to μ''_s.

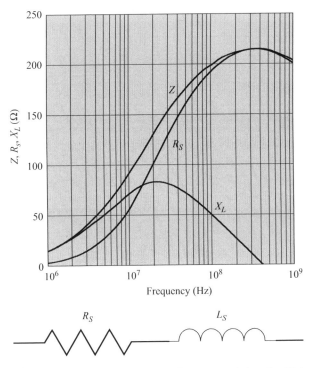

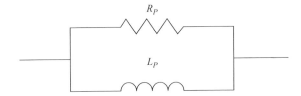

Figure 2-2   Another representation of complex permeability for Fair-rite material 61. These data present the same information as Figure 2-1 but as an equivalent resistance, reactance, and total impedance.

Figure 2-3   Parallel transformation of the schematic in Figure 2-2. R_P will be very large compared with R_S.

An alternate equivalent circuit can be derived via a series of parallel conversions (Figure 2-3). A more complete model of the impedance includes two parallel resonant circuits in series and is shown in Figure 2-4. The capacitor (C_c) is the parasitic capacitance of the winding of the inductor. This capacitance can be quite large in an MnZn core where relative permittivity is on the order of 10^4. L_C is the inductance due to the winding on the core, and R_C is the core loss due to hysteresis. An NiZn core will have less capacitance since relative permittivity varies from 100 to 10 over frequency. The permittivity parallels the frequency dependence of MnZn

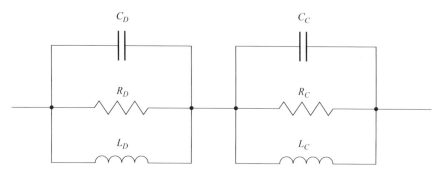

Figure 2-4   A more complete model of an inductor wound on a ferrite core. The items with a "D" subscript are due to dimensional resonance. The items with a C subscript are due to the coil of wire wound on the core.

magnetic properties where the useful frequency range ends around 10 MHz. The larger effective capacitance changes the impedance from inductive to capacitive at a relatively low frequency.

The second parallel circuit in Figure 2-4 is caused by dimensional resonance, which E. C. Snelling described in [7–9]. Dimensional resonance is exactly the same as that in a cavity resonator or dielectric resonator. MnZn materials have very high relative permeability and relative permittivity, which combine to make the wave velocity inside the material very low. When one dimension of the core is one-half wavelength and perpendicular to the magnetic field, the core will support a standing wave. This completely cancels the magnetic flux, and the apparent permeability drops to zero. The magnetic loss peaks at resonance. If the material has high permeability losses, it will not support standing waves but the magnetic flux will also only partially penetrate the core. NiZn materials have much lower permeability and permittivity compared with MnZn materials, so the dimensional resonance tends to be near 1 GHz in most NiZn cores. L_D and C_D are the equivalent values that relate to the resonator dimensions. R_D is due to the loss of the material and is caused principally by eddy current losses with a minor contribution from hysteresis loss.

2.4   Magnetic Losses

The aforementioned loss mechanisms primarily apply to low-level operation. For higher power operation, a more appropriate presentation is the power loss density that expresses conversion of the applied field to heat loss per unit volume. Figure 2-5 is a representative graph of power loss versus both magnetic flux density and frequency for the MnZn material 3F5. Such graphs are not provided by any of the manufacturers of NiZn materials. This is unfortunate when we wish to create a high frequency transformer using either magnetic or transmission line modes.

Snelling identified the losses in a ferrite as hysteresis loss, eddy current loss, dimensional resonance, and residual losses (relaxation loss, ferromagnetic resonance loss, domain wall loss, thermal after effect loss).

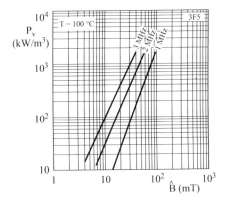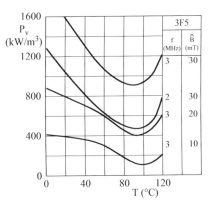

Figure 2-5   *Power loss density data for Ferroxcube 3F5 material shows loss versus frequency, volume, temperature, and flux density. This material becomes more efficient as the temperature rises. Loss curves are representative of both MnZn and NiZn materials.*

Hysteresis losses are caused by the irreversible movement of domain walls and irreversible domain rotations. The energy lost during an entire cycle of applied AC voltage is equal to the hysteresis loop area. As long as the applied field is relatively small, the hysteresis loop does not tilt appreciably. As the field increases, the area of the loop increases and the curve tilts closer to saturation. The tilt toward saturation is responsible for creating harmonic distortion in the circuit. Harmonic distortion is a real consideration in transmission line transformers in the low frequency region where we rely on magnetic coupling.

Eddy current losses arise because of the alternating flux in the conducting material. The changing magnetic field induces current flow in the material, which becomes heat due to I^2R loss. Eddy current losses increase with the square of frequency because the current loops decrease in diameter with increasing frequency. At lower frequencies the insulating grain boundaries inhibit current flow because the loops are large compared with the size of the grains. Eddy currents increase significantly when the eddy current loops approach the grain size.

A significant residual loss is caused by ferromagnetic resonance. The electrons in the magnetic material act as very small gyroscopes. The applied field causes the electron spin to precess. As frequency increases the field begins to approach the natural frequency of the precession. This causes a dispersion of the permeability and losses increase near resonance. The high frequency limit of a material is due to this resonance.

Early experiments by Sevick indicated that the bulk resistivity of ferrite material could be related to high efficiency operation. Therefore, many of the major suppliers were asked to supply samples of their highest resistivity material. Table 2-3 is a list of suppliers that provided samples to Sevick, the code symbol for their materials, the low frequency (initial) permeability, and the bulk resistivity. Powdered iron was included because it has been used for some applications, but as

Table 2-3   Cores, Suppliers, and Specifications

Material	Supplier	Permeability	Bulk Resistivity (Ω-cm)
Q1 (NiZn)	Allen-Bradley (formerly Indiana General)	125	10^8
G (NiZn)	Allen-Bradley	300	10^6
Q2 (NiZn)	Allen-Bradley	40	10^9
H (NiZn)	Allen-Bradley	850	$10^4 - 10^5$
4C4 (NiZn)	Ferroxcube	125	$10^7 - 10^8$
3C8 (MnZn)	Ferroxcube	2700	$10^2 - 10^3$
K5 (NiZn)	MH&W Intl (TDK)	290	2×108
KR6 (NiZn)	MH&W Intl (TDK)	2000	$10^5 - 10^6$
CMD5005 (NiZn)	Ceramic Magnetics	1400	7×10^9
C2025 (NiZn)	Ceramic Magnetics	175	5×10^6
CN20 (NiZn)	Ceramic Magnetics	800	10^6
C2050 (NiZn)	Ceramic Magnetics	100	3×10^7
E (powdered iron)	Arnold Engineering, Amidon Associates	10	10^{-2}

will be shown, it suffers by comparison because of its very low permeability. Many of the original manufacturers are no longer in business or have been absorbed by other companies. In the United States, only Ferroxcube (now part of Yageo), TDK (no longer supplying NiZn material), and Ceramic Magnetics remain from the original list as suppliers of NiZn materials suitable for high frequency transformer use. Table 2-5 gives an up-to-date (as of 2013) list of suppliers and distributors.

All of the data Sevick presents in this book on loss as a function of frequency were obtained at Bell Labs on a computer-operated transmission measuring set with an accuracy of 0.001–0.002 dB over a frequency range of 50 to 1000 MHz [10–12]. As a reference, a loss of 0.044 dB represents a loss of 1%, or an efficiency of 99%. Actual data show that many of the transformers, made using the best ferrite materials, exhibit losses over a considerable portion of their passband of only 0.020 dB, equivalent to efficiencies of 99.5%. Since short windings (wire lengths of 10 to 15 in) are generally used, very little loss is attributed to the windings. In fact, wires as thin as no. 18 can easily handle 1000 W. Although later chapters stress theory and design, many of the experimental results presented do display the high efficiency of the transmission line transformer.

2.5   Ferrites and Frequency Response

In later sections, Sevick demonstrates that the core still performs a major role at the high frequency end of a transmission line transformer. The question frequently asked is, how can a ferrite material like Q1, 4C4, or 61 designed for operation up to 10 or 20 MHz still produce a flat response beyond 100 MHz? The answer lies in Figure 2-6, which shows that the typical response of most ferrites approaches the same permeability value at the high frequency end. Since the reactance of a coiled (or beaded) transmission line is proportional to the product of the frequency and

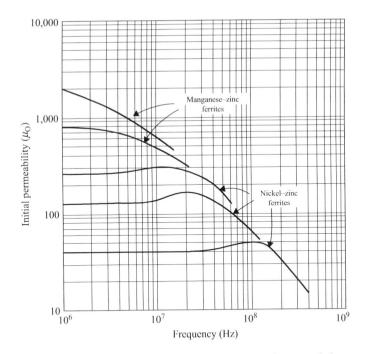

Figure 2-6   The curves for various materials show initial permeability versus frequency for various ferrite materials. The downward slope of all curves approaches the same slope. The decrease in permeability is roughly the same as the increase in frequency, so the product of μ times frequency is a constant.

permeability and the slope beyond the knee of the curves in Figure 2-6 shows a near-constant product of permeability and frequency, the following applies:

1. The maximum reactance of the winding for each ferrite occurs just beyond the knee of the curve.
2. It has a constant value with frequency beyond the knee of the curve.
3. Surprisingly, the maximum values are about the same for all ferrites.

Then why not use the highest permeability ferrite? Fewer turns would be needed at the low frequency end, which would in turn enhance the high frequency end (particularly with Ruthroff transformers). In power applications, only nickel-zinc ferrites with permeability below 300 produce efficiencies in excess of 98%. High permeability materials, like manganese-zinc ferrite, do not produce good efficiency and are not recommended for power applications.

2.6   Power Ratings

Power ratings are generally determined by two conditions: (1) temperature rise due to losses; and (2) exceeding the maximum values of operating parameters by accident. A failure caused by an increase in temperature is usually time-dependent, while the breakdown of a device operated over its maximum ratings is instantaneous. NiZn materials, in general, are sensitive to excessive magnetic flux and mechanical stresses. Many material sheets warn against subjecting cores to these excessive stresses.

Losses in a magnetic transformer are caused by the loss mechanisms shown in section 2.3. In general, the insulators in a transformer will fail due to excessive temperature before the core will fail due to loss of permeability (Curie temperature). Figure 2-5 shows loss curves for Ferroxcube material 3F5, which is representative of MnZn materials. We would expect similar curves for NiZn materials if manufacturers did such measurements. Of course, the frequencies would be much higher. The heat loss actually decreases with increasing temperature until the core is near 100°C and then increases with increasing temperature until the Curie temperature is reached. For 3F5 material, the Curie temperature is 325°C, so it is likely that any soldered connections would fail long before the permeability reaches zero. It is also instructive that the loss is an exponential function of both flux density and loss per unit volume as shown with a log-log plot. Tripling the flux density causes a tenfold increase in heat loss for this material.

Different conditions exist when a transmission line transformer's power rating is determined. Because of the canceling effect of the transmission line currents, little flux is generated in the core. This holds true even when tapped multiwindings are involved. Since losses with certain ferrites are only on the order of 0.02–0.04 dB, very small transmission line transformers can handle surprisingly high power levels. Losses in ferrite materials increase with impedance levels because we usually have more volts per turn, and these levels must be considered when designing a transmission line transformer. Most failures in transmission line transformers are of the catastrophic type and are usually caused by poorly terminated (or unterminated) transformers. Such conditions create high voltages and a breakdown in the insulation between the windings. This is particularly true of close-wound enameled or Formvar-type wires. Enamel or Formvar wire typically has insulation rated at 6000 V. An unterminated winding can exceed such a voltage.

Standards for setting power ratings for transmission line transformers have not appeared in the literature, nor are they available from any of the suppliers of ferrite material. The data presented in this book appear to be the only quantitative information available on the losses of these transformers. Because limited reliability information is available on transmission line transformers, and losses generally increase with impedance levels, an exact formulation of power ratings is difficult to make. But as a result of Sevick's findings, some general guidelines can be offered when considering ratings:

1. The power capability of these devices (when energy is transmitted from input to output by transmission line mode) is determined more by the size of the conductors and not by the cores. Very small structures can handle amazingly high power levels. Thus, larger wires or the use of coaxial cable or flat parallel line can more than double the power ratings.

2. The voltage ratings can be increased significantly by the use of polyimide or polyamide-coated wires. In many cases, to optimize the characteristic impedance of the windings, extra layers of 3M no. 92 tape or DuPont Kapton (another polyimide insulation) are used. This also increases the breakdown voltage.

3. Generally, the lowest-permeability NiZn ferrites yield the highest efficiencies. These have permeability in the range of 40 to 50 but can limit the low frequency response. When operating at impedance levels below 100 Ω, permeability as high as 300 should yield very high efficiencies (98 to 99%). When operating at impedances above 100 Ω, the trade-off is low frequency response for efficiency. Actually, most of the ferrites with permeability of 200 to 300 can still yield acceptable efficiencies (at least 97%) at the 200 to 300 Ω impedance level.

4. Very few differences in efficiency were observed from the ferrites supplied by the manufacturers listed. Limited measurements on 4C4 material ($\mu = 125$) from Ferroxcube showed the best efficiency at the 20 Ω level. This material is also reported to be free of the failure mechanism due to high flux density exhibited by most of the other NiZn ferrites.

5. Although many examples in this text refer to the various company designations for the ferrites used, practically any other ferrite (with the same permeability) can be substituted.

6. When transformers become warm to the touch (after the power is turned off!), it suggests that either the wrong ferrite is used or that the reactance of the coiled windings, at the frequency in question, is insufficient to prevent conventional transformer currents. The problem is probably not in the size of the conductors.

Table 2-4 lists some suggested power ratings. Although efficiencies can vary with permeability and impedance level, these ratings should generally hold for permeability below 300. Also, since the currents can vary with impedance level and the position of the winding in a higher-order winding, the transmission line descriptions are offered only as a general practice.

Several small transformers were tested under severe conditions to check the validity of the ratings in Table 2-4. One 4:1 transformer had 10 turns of no. 18 wire

Table 2-4   Suggested Power Ratings for Ferrites with Permeability Below 300

Core Size	Description of Transmission Line	Rating (Continuous Power)
1 in OD toroid, 1/4 in diameter rod	16–18 gauge wire	200 W
1 1/2 in OD (or greater) toroid, 3/8 in diameter (or greater) rod	14 gauge wire	1000 W
1 1/2 in OD (or greater) toroid, 3/8 in diameter (or greater) rod	10–12 gauge wire, coaxial cable, flat parallel transmission line	2000 W

Table 2-5   Material Suppliers

General Distributors

Amidon Associates, Inc.
240 Briggs Ave
Costa Mesa, CA 92626
Phone: 714-850-4660
Toll Free: 1-800-898-1883
Fax: 714-850-1163
Email: sales@amidoncorp.com

Components:
Fair-Rite NiZn toroids, binocular cores, rods, beads, engineering kits
Powdered Iron cores
Other material manufacturers: Tokin, Ferroxcube, Magnetics, Epcos
Magnet wire
Polyimide tape
PTFE tubing

The Wireman
261 Pittman Rd
Landrum, SC 29356
Phone: 800-727-9473

Components:
NiZn beads
Iron powder toroids
Polyimide tape
Magnet wire

NiZn Ferrite Manufacturers

Ceramic Magnetics
16 Law Drive Fairfield, NJ 07004
Phone: 973-227-4222
Fax: 973-227-6735
Website: www.cmi-ferrite.com
Distributors: None, but sells direct

Fair-Rite Products Corp
Box J
Wallkill, NY 12589
Phone: 914-895-2055
Website: www.fair-rite.com
Distributors: Amidon, Allied, Dexter, Mouser, Newark, Elna

Yageo (Ferroxcube)
Regional Ferroxcube sales office, El Paso, TX
Phone: 915-599-2328
Fax: 915-599-2555
Website: www.ferroxcube.com
Distributors: Amidon, Elna, Adams, Allstar

Table 2-5   (Continued)

MMG Canada Limited
10 Vansco Road
Toronto, Ontario M8Z 5J4
Canada
Phone: +1-416-251-2831
Fax: +1-416-251-6790
Distributors: Adams, Allstar

Wire and Insulation Manufacturers

Belden
401 Pennsylvania Parkway, #200
Indianapolis, IN
Phone: 317-818-6300
Fax: 317-818-6382

Phelps Dodge
Magnet Wire Company
2131 South Coliseum Blvd
Fort Wayne, IN 46801
Phone: 260-421-5400
Fax: 260-421-5412
Wire: H Imideze

Consolidated Electronic Wire & Cable
11044 King St.
Franklin Park, IL 60131
Phone: (toll-free) 800-621-4278 ext. 8226 (local) 847-455-8830
Fax: 847-455-8837
Website: www.conwire.com

3M Company
Industrial Electrical Products Division
3130 Lexington Ave. S
Eagan, MN 55121
Phone: 800-233-3636
Insulation: Scotch nos. 27, 92

Magnetic Core Distributors

Adams Magnetic Products Co.
888 Larch Avenue
Elmhurst, IL 60126
Phone: 630-617-8880
Fax: 630-617-8881
Website: www.adamsmagnetic.com

Allstar Magnetics
6205 NE 63rd Street
Vancouver, WA 98661

(Continues)

Table 2-5   (Continued)

Phone: 360-693-0213
Fax: 360-693-0639
Website: allstarmagnetics.com

Amidon Associates, Inc.
240 Briggs Ave
Costa Mesa, CA 92626
Phone: 714-850-4660
Toll Free: 1-800-898-1883
Fax: 714-850-1163
Website: www.amidoncorp.com

Dexter Magnetic Technologies, Inc.
1050 Morse Avenue
Elk Grove Village, IL 60007-5110
Phone: (toll-free) 800-345-4082 (local) 847-956-1140
Fax: 847-956-8205
Website: www.dextermag.com

Elna Magnetics
203 Malden Turnpike
Saugerties, NY 12477
Phone: (toll-free) 800-553-2870 or 800-223-3850 (local) 845-247-2000
Website: www.elnamagnetics.com

MH&W International Corp.
14 Leighton Place
Mahwah, NJ 07430-3119
Phone: 201-891-8800
Website: www.mhw-intl.com

on a Q1 toroid with a 1 in OD. The other transformer had 14 turns of no. 18 wire on a Q1 rod with a 1/4 in diameter. These transformers, operating at an impedance level of 50:12.5 Ω, successfully handled 1 kW of peak power in single-sideband operation over an extended period of time. They became warm to the touch but showed no evidence of damage.

Accurate loss measurements have shown that only a limited number of ferrite materials are useful in power applications, where high efficiency is an important consideration. This chapter describes transmission line transformers that use nickel-zinc ferrite cores with permeability in the moderately low range of approximately 50 to 300 to yield efficiencies in excess of 98%. No conventional transformer can approach this performance. The losses are not a function of current as in the conventional transformer but are, in most cases, related to the impedance levels at which the transformers are operated. This suggests a dielectric loss rather than the conventional magnetic loss caused by core flux.

2.7 Suppliers of Materials

Table 2-5 lists materials manufacturers and distributors. The easiest method of obtaining cores and rods in small quantities is to contact the appropriate distributor. In general, manufacturers are not set up for small orders. Fair-Rite, MMG Canada, and Ceramic Magnetics have been especially supportive with data and materials for use in updating the information for the current edition.

2.8 Additional Reading

Numerous Internet resources such as catalogs, data sheets, and application notes from distributors and manufacturers were not available when Sevick wrote the first few editions of this book and are a major change over the past 30 years. As of the time of this writing (2013), the best place to find useful application information is on the website of the distributors of magnetic materials. For example, the power density graphs for materials 43 and 61 on Amidon's site are the only presentations of power density available for any NiZn materials.

This chapter has examined dimensional resonance. Additional reading on this topic can be found in Jim Brown's "A Ham's Guide to RFI, Ferrites, Baluns, and Audio Interfacing" [13]. He provides experimental data showing the effects of distributed capacitance, inductor turns, and dimensional resonance for materials 43, 61, and 78 and also gives some excellent analysis of the relative merit of various balun configurations.

References

[1] Takei, T., "Review of Ferrite Memory Materials in Japan," Ferrites, *Proceedings of the International Conference*, ed. Y. Hoshimo, S. Jida, and M. Sugimoto, Baltimore, MD: University Park Press, pp. 436–437.

[2] Snoek, J. L., *New Developments in Ferromagnetic Materials*, New York: Elsevier, 1947.

[3] Stone Jr., H. A., "Ferrite Core Inductors," *Bell System Tech Journal*, Vol. 32, Mar. 1953, pp. 265–291.

[4] Slick, P. I., US Pat No. 3,533,949; filed Nov. 21, 1967, issued Oct. 13, 1970.

[5] Guanella, G., "Novel Matching Systems for High Frequencies," *Brown-Boveri Review*, Vol. 31, Sep. 1944, pp. 327–329.

[6] Ruthroff, C. L., "Some Broad-Band Transformers," *Proceedings of the IRE*, Vol. 47, Aug. 1959, pp. 1337–1342.

[7] Snelling, E. C., *Soft Ferrites, Properties and Applications*, Chemical Rubber Publishing, 1969.

[8] Snelling, E. C., and A. D. Giles, *Ferrites for Inductors and Transformers*, Research Studies Press, 1983.

[9] Snelling, E. C., *Soft Ferrites, Properties and Applications*, 2nd ed., Butterworth-Heinemann, 1989.

[10]   Geldart, W. J., G. D. Haynie, and R. G. Schleich, "A 50-Hz –250-Mhz Computer Operated Transmission Measuring Set," *Bell Systems Tech Journal*, Vol. 48, No. 5, May–Jun. 1969.

[11]   Geldart, W. J., and G. W. Pentico, "Accuracy Verification and Inter-comparison of Computer-Operated Transmission Measuring Sets," *IEEE Transactions on Instruments and Measurement*, Vol. IM-21, No. 4, Nov. 1972, pp. 528–532.

[12]   Geldart, W. J., "Improved Impedance Measuring Accuracy with Computer-Operated Transmission Measuring Sets," *IEEE Transactions on Instruments and Measurement*, Vol. IM-24, No. 4, Dec. 1975, pp. 327–331.

[13]   Brown, J., "A Ham's Guide to RFI, Ferrites, Baluns, and Audio Interfacing," June 5, 2010, http://audiosystemsgroup.com/RFI-Ham.pdf

Chapter 3

Guanella Analysis

3.1   Introduction

The transmission line transformers Guanella presented in 1944 were high impedance open wire transmission lines wound on insulating forms. Coiling the transmission line made the transformer physically smaller and reduced common mode current flow. Probably the largest commercial use of his design was the 75:300 Ω balun used in the VHF tuner of TV receivers, especially by RCA, in the 1950s and 1960s.

Guanella's transformers are an example of equal delay transformers. The delay through all of the component transmission lines is the same. The transformers operate by adding or subtracting voltages and currents in phase. His investigation was directed toward developing a broadband transformer for matching the balanced output of a 100 W push–pull vacuum tube amplifier to the unbalanced load of a coaxial cable. The objective was to match a balanced impedance of 960 Ω to an unbalanced impedance of 60 Ω (16:1) from 100 to 200 MHz. His experimental data, with a 53 Ω resistor as a load, showed a deviation of less than 10% from the theoretical value over this frequency range. Guanella accomplished this by incorporating four 240 Ω transmission lines in a parallel–series arrangement resulting in a high impedance, 16:1 balun. Sevick (and probably others) had overlooked his technique of summing in-phase voltages at the high impedance side of the transformer, as is evidenced by the scarcity of information on Gunaella's designs in the literature. This chapter presents the analysis of his transformers and explains the advantages of his technique with transformers for high power and high impedance applications.

3.2   Mid-Band Operation

Figure 3-1 is a schematic of Guanella's 1:4 transformer. The two transmission lines are in parallel on the low impedance side and in series on the high impedance side. With the single connection to ground the transformer performs as a step-up balun with a floating load.

The mid-band performance is determined by the characteristic impedance of the transmission lines. For flat lines, the desired characteristic impedance is $Z_0 = R_L/2$. When parasitic elements are absorbed into the characteristic impedance, this transformer (as Guanella stated in his paper) yields a frequency-independent

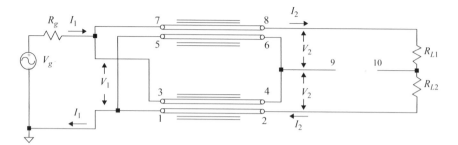

Figure 3-1   Schematic shows the Guanella 1:4 transformer. The symbol for each parallel transmission line includes the ferrite loading to extend the frequency range.

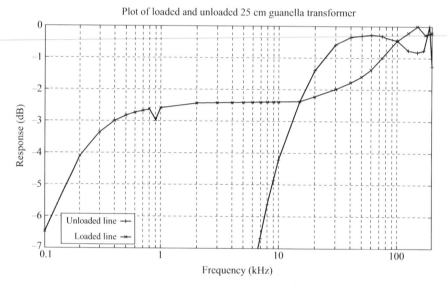

Figure 3-2   The curves show the frequency response of a 25 cm Guanella transformer composed of 0.141 semi-rigid coax. The response is distorted due to the grounds of the oscilloscope used for measuring the response. However, the response of the unloaded transformer drops off significantly below 20 MHz. The response of the transformer loaded with material 43 beads is flat from approximately 200 kHz to 200 MHz.

transformation. The transformer operation will work at all frequencies where the length of the transmission line ensures transverse electromagnetic (TEM) operation.

Figure 3-2 shows the frequency response of a 25 to 100 Ω transformer composed of two pieces of 0.141 diameter 50 Ω cable that are 25 cm long.

The transmission lines begin to lose their ability to transform near 10 MHz where the line length is approximately 0.03 λ. Additionally, the stray impedances (especially lead inductance of the 100 Ω load) cause an increase in impedance at high frequencies. The transformed impedance drops quickly below 10 MHz. The high frequency limit is controlled entirely by the losses in the coax and stray effects. The second transformer is loaded with three Fair-Rite material 43 cores (#2643626502) on each line. Notice that the cores perform as Guanella predicted. The effect of stray elements is drastically reduced over most of the range, and the lower range extends down a full decade. Slightly degraded performance is extended almost two full decades.

It might not be obvious how a coaxial cable can be both the primary and secondary of the transformer at lower frequencies since the shield completely encloses the inner conductor. It is, in fact, a Faraday shield and completely encloses the electric field between the two conductors. Coaxial cable is produced using either copper or aluminum for the shield. Both of those materials have magnetic permeability of approximately -1.0×10^{-5}, so the magnetic field from the inner conductor passes through the shield and interacts with the magnetic material of the core. Obviously, we would not want to make our coax from a magnetic material such as silver-plated steel since it would make an excellent magnetic and electric shield. The transition from transmission line mode to magnetic mode is more intuitive with parallel wire transmission line wound on a toroid. In that case, the transmission line becomes just two wires wound on a core with the windings closely wound next to each other.

With two transmission lines, as in Figure 3-1, the input impedance at the low side is

$$Z_{in} = \frac{1}{2} \times Z_0 \left(\frac{\frac{Z_L}{2} + jZ_0 \tan \beta L}{Z_0 + j\frac{Z_L}{2} \tan \beta L} \right) \tag{3-1}$$

where

$Z_0 =$ the characteristic impedance
$Z_L =$ the load impedance
$L =$ the length of the transmission line in wavelengths
$\beta = 2\pi/\lambda$, where λ is the the effective wavelength in the transmission line

With the optimum value of $Z_0 = R_L/2$ for a resistive load, equation (3-1) reduces to

$$Z_{in} = R_L/4 \tag{3-2}$$

With more than two transmission lines, equation (3-2) becomes

$$Z_{in} = R_L/n^2 \tag{3-3}$$

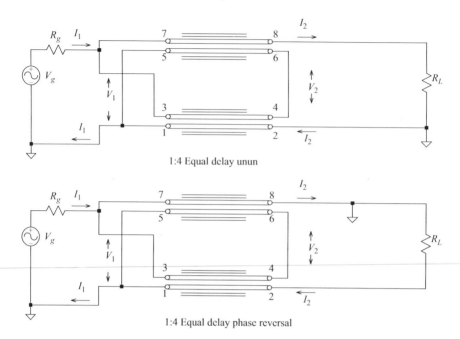

Figure 3-3   *These diagrams show representative equal-delay unun and phase reversal transformers using Guanella's method.*

where n is the number of transmission lines. Conversely, upon inspection, when looking in at the high impedance side,

$$Z_{in} = n^2 R_L \tag{3-4}$$

where R_L would be the low impedance on the left side in Figure 3-1.

Placing two ground connections can produce other transformers such as unbalanced-to-unbalanced transformers (*ununs*), phase reversals, and hybrids. Figure 3-3 shows representative equal delay unun and phase reversal circuits.

3.3 Low Frequency Operation

Guanella did not have the advantage of modern magnetic materials to extend his concept to even lower frequencies. The subsequent discussion still follows the concepts of Guanella but extends the frequency significantly lower.

One low frequency model of the Guanella 1:4 transformer is shown in Figure 3-4. It represents the case where energy is no longer transmitted from input to output by transmission line mode. This step in the evolution of the Guanella transformer loads each transmission line with ferrite magnetic material.

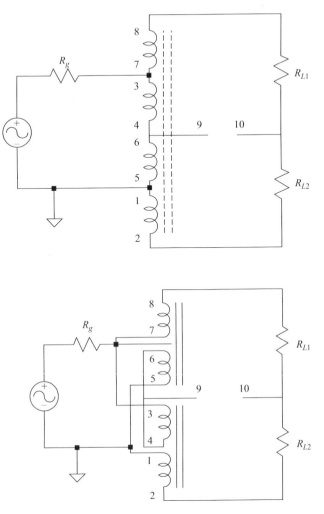

Figure 3-4 The graphs show the low frequency equivalent of two Guanella 1:4 transformers: (Top) Equivalent with both lines wound on the same core. (Bottom) Circuit with each line wound on its own core.

Because the transmission line is no longer operating in TEM mode, each transmission line now acts entirely as a magnetically coupled autotransformer. The combined circuit is nothing more than two 1:1 transformers with the primary windings in series and their secondary windings set up as autotransformers in series aiding. The combination results in a 2:4 turns ratio autotransformer yielding a 1:4 impedance ratio.

The low frequency circuit is similar if both cables are part of the same magnetic circuit, but the interactions in the transition from transmission line to magnetic operation may be slightly different. Once again, the transformer has two windings in series on the input side and two windings as autotransformers, resulting in a 2:4 autotransformer. The difference is that there is now more capacitive coupling between the coax shields as well as capacitance between each center conductor and its associated shield. If the two windings are on separate cores, the magnetizing inductance is just the sum of the two separate inductances. But if a single core is used and the windings are in the same direction to be series aiding, then the total magnetizing inductance would be greater by a factor of two.

3.4   Guanella Transformer Variations

Figure 3-5 shows variations on the 1:4 transformer, which we have already seen used as a step-up balun. The device is bilateral, so the configuration can just as easily be used as a 4:1 step-down balun by connecting a grounded source to the high impedance side and taking the balanced or floating load from the low

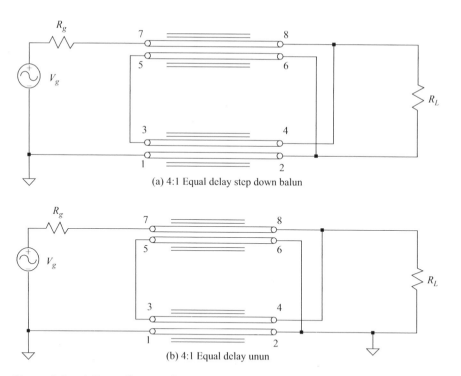

(a) 4:1 Equal delay step down balun

(b) 4:1 Equal delay unun

Figure 3-5   A Guanella transformer can be used for step-down operation as well as step-up operation. (a) Step-down version of equal-delay balun. (b) Step-down version of equal-delay unun.

impedance side (Figure 3-5a). Whether the ground and generator are on the low impedance side in a step-up balun or the ground and generator are on the high impedance side in a step-down balun, the low frequency response is the same.

Figure 3-5b shows that the 1:4 transformer can be used as an unun transformer. However, this operation is more problematic at lower frequencies. The low frequency response is highly dependent on where the ground connections are made and whether one or two cores (or beaded lines) are employed. As soon as the magnetizing inductance at a given frequency becomes low enough, the line grounded at both ends will short out the transformer. We will look at alternatives in later chapters.

Figure 3-6a shows an equal-delay 1:1 balun, and Figure 3-6b shows a phase reversal transformer. They have the same issue with low frequency response (Figure 3-5b). Once the frequency is low enough the magnetizing inductance no longer keeps the output isolated from the input.

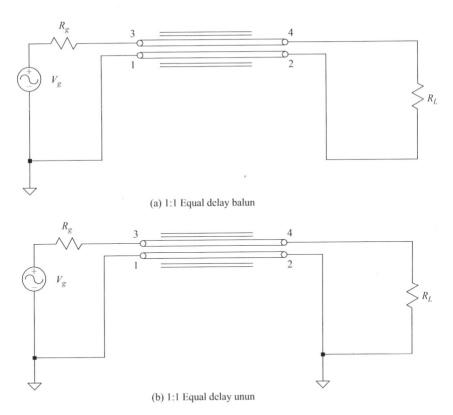

(a) 1:1 Equal delay balun

(b) 1:1 Equal delay unun

Figure 3-6   Guanella's method can be used for a 1:1 equal delay (a) balun and (b) unun.

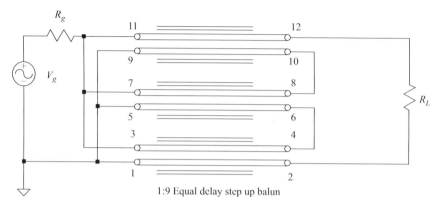

1:9 Equal delay step up balun

Figure 3-7   This diagram illustrates a 1:9 equal-delay step-up balun. If the lines have 150 Ω characteristic impedance, it will match 50 to 450 Ω.

Figure 3-7 shows a three-line transformer that yields a 1:9 transformer for 50:450 operation. In theory, this technique can be extended to any number of transmission lines resulting in any $1:n^2$ result desired. We will look at very low impedance as well as very high impedance issues in later chapters. Guanella connected all transmission lines in parallel at one end of the transformer and all of the far end transmission lines in series. We will also look at noninteger combinations of series and parallel combinations where there is a mix of series and parallel connections at both ends of the transformer. Such combinations make ratios such as 1:2.25 possible.

Chapter 4

Ruthroff Analysis

4.1   Introduction

Ruthroff presented an alternative technique to Guanella for obtaining a 1:4 impedance transformation in his classic 1959 paper. His concept involved summing a direct voltage with a delayed voltage that traversed a single transmission line. Since his investigations involved small signal applications, he was able to use very small, high permeability cores and fine wires. His manganese-zinc cores ranged only from 0.175 to 0.25 in diameter and from 1600 to 3000 in permeability. His conductors, which were twisted to control the characteristic impedance, were only AWG 37 and 38 wires. Since the transmission lines were very short under these conditions (therefore little phase shift between the summed voltages), he was able to demonstrate pass bands essentially "flat" from 500 kHz to 100 MHz.

Guanella and Ruthroff both designed transformers that combine transmission line operation with magnetic transformer operation. However, Ruthroff designs operate in magnetic mode over a significantly larger portion of the effective bandwidth. His designs use very short transmission lines to extend the high frequency end of operation. Heaviside showed that a transmission line ceases to act as a transmission line when the length is less than approximately 0.1 wavelength. We will see shortly that a Ruthroff design is useful up to 0.25 wavelength when the load is not optimum or near 0.5 wavelength if the load is optimum. This implies that a significant portion of the useful band will be near or below the 0.1 wavelength transition region.

Figure 4-1 shows the high frequency schematics of the two 1:4 transformers presented by Ruthroff. Figure 4-1a has the basic building block in the bootstrap configuration, which results in a 1:4 unbalanced-to-unbalanced (unun) transformer. Figure 4-1b has the basic building block in the phase reversal configuration, which results in a 1:4 balun. These high frequency models assume sufficient longitudinal reactance of the windings such that the outputs are completely isolated from the inputs. Unlike Guanella's model, which can be analyzed by inspection, Ruthroff resorted to loop and transmission line equations to solve for the power in the load and hence transducer (insertion) loss.

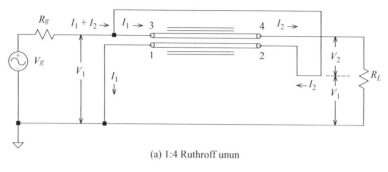

(a) 1:4 Ruthroff unun

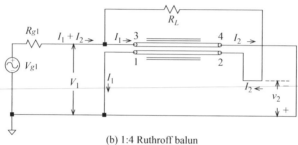

(b) 1:4 Ruthroff balun

Figure 4-1   Schematics show a Ruthroff transformer example of (a) a 1:4 unun and (b) a 1:4 balun.

The ununs in Figure 4-1a are as follows:

$$V_g = (I_1 + I_2)R_g + V_1$$
$$I_2 R_L = V_1 + V_2$$
$$V_1 = V_2 \cos \beta l + j I_2 Z_0 \sin \beta l$$
$$I_1 = I_2 \cos \beta l + j V_2 / Z_0 \sin \beta l$$

(4-1)

where β is the phase constant of the line (velocity factor), and l is the length of the line in radians. Ruthroff found that the maximum transfer of power occurs when $R_L = 4R_g$ and that the optimum value of the characteristic impedance is $Z_0 = 2R_g$. Figure 4-2 shows the loss as a function of the normalized line length and for various values of the characteristic impedance, Z_0. Even with the optimum value of the characteristic impedance, the loss is found to be 1 dB when the line is a quarter wavelength and infinite when it is a half wavelength. Figure 4-2 illustrates the value of keeping the transmission line as short as possible with Ruthroff's 1:4 unun.

Ruthroff also derived equations for the input impedances seen at either end of the transformer with the opposite end terminated in Z_L. They are:

$$Z_{in}(\text{low impedance end}) = Z_0 \left[\frac{Z_L \cos \beta l + j Z_0 \sin \beta l}{2Z_0(1 + \cos \beta l) + j Z_L \sin \beta l} \right]$$

(4-2)

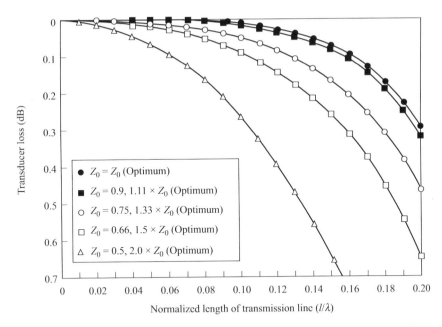

Figure 4-2   *The curves show loss as a function of normalized transmission line length in a Ruthroff 1:4 unun for various values of characteristic impedance, Z_0.*

and

$$Z_{in}(\text{High impedance end}) = Z_0 \left[\frac{2Z_L(1 + \cos \beta l) + jZ_0 \sin \beta l}{Z_0 \cos \beta l + jZ_L \sin \beta l} \right] \tag{4-3}$$

Pitzalis, *et al*, plotted Z_{in} (low impedance end) as a function of various values of Z_0 compared to the optimum value $Z_0 = 2R_g$. Figures 4-3 and 4-4 are reproductions of their curves for the real and imaginary parts of the input impedance. These curves can be returned to impedance by multiplying the ordinate value by $R_L/4$.

They found that the input impedances were also sensitive to the value of the characteristic impedance. Looking into the low impedance side of the transformer, the following concepts can be generalized:

1. For a Z_0 greater than the optimum value:

    (a) The real part of Z_{in} increases only slightly with increasing frequency and values of Z_0.

    (b) The imaginary part of Z_{in} becomes positive and increases with frequency and values of Z_0.

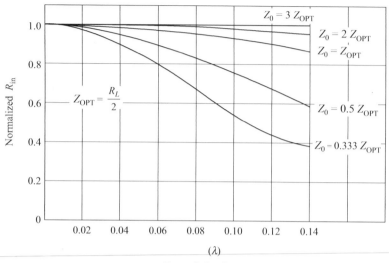

Figure 4-3   *The normalized real part of the input impedance of a Ruthroff 1:4 unun is shown as a function of Z_0 and the length of the transmission line.*

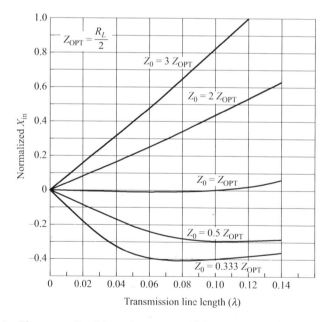

Figure 4-4   *The normalized imaginary part of the input impedance of a Ruthroff 1:4 unun is shown as a function of Z_0 and the length of the transmission line.*

2. For a Z_0 less than the optimum value:

   (a) The real part of Z_{in} decreases greatly with increasing frequency and decreasing values of Z_0.

   (b) The imaginary part of Z_{in} becomes negative and increases in magnitude with frequency and values of Z_0.

The high frequency model of Ruthroff's 1:4 balun (Figure 4-1b) adds a direct voltage V_1 to a delayed voltage $-V_2$ as in the basic building block connected as a phase inverter. It can be shown that the high and low frequency responses are the same as his 1:4 unun. Two other comments can be made regarding this approach to a 1:4 balun. They are:

1. Unlike Guanella's balun, this one is unilateral; that is, the high impedance side is always the balanced side.

2. When the center of the load, R_L, is grounded, the high frequency response is greatly improved. The balun now performs as a Guanella balun which sums two in-phase voltages.

4.2 Low Frequency Analyses of Ruthroff's 1:4 Transformers

Figure 4-5 shows the low frequency models for the two Ruthroff 1:4 transformers. Figure 4-5a is the schematic for his 1:4 unun, and Figure 4-5b for his 1:4 balun. These models represent the cases when the longitudinal reactance of the coiled transmission lines are insufficient and energy is no longer transmitted by a transmission line mode. Figure 4-5a can be recognized as the schematic of a 1:4 autotransformer. Although the analysis presented here is for the 1:4 unun, it can be shown that the 1:4 balun has the same result.

As with the conventional autotransformer, the low frequency performance of the Ruthroff 1:4 unun can be determined from the reduced model shown in

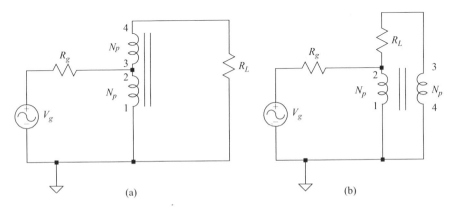

Figure 4-5 *Schematics of the low frequency model of Ruthroff 1:4 transformers (a) unun and (b) balun show the similarity to an autotransformer.*

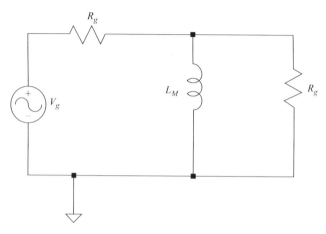

*Figure 4-6   The diagram illustrates the low frequency electrical model for a
1:4 autotransformer. The inductor in parallel with the transformed
load (the second R_g element) explains the reduction in response as
frequency decreases.*

Figure 4-6. Here we have an ideal transformer, shown by the load, labeled R_g,
shunted by the core magnetizing inductance, L_M.

If a toroid is used for the core, the magnetizing inductance (L_M) is

$$L_M = 0.4\pi N_p^2 \mu_0 \left[\frac{A_e}{l_e}\right] \times 10^{-8} \, henrys \tag{4-4}$$

where

N_p = the number of primary turns
μ_0 = the permeability of the core
A_e = the effective cross sectional area of the core in cm^2
l_e = the average magnetic path length in the core in cm

Equation (4-4) has some important features. By making the outside diameter of
the toroid as small as is practical while keeping the same cross-sectional area,
a large improvement in bandwidth takes place at both the low and high frequency
ends of the transformer. With smaller toroids, the length of the transmission line is
shorter and the magnetizing inductance is larger because of the shortened average
magnetic path length. By using the highest permeability consistent with high efficiency, the low frequency response is helped further. In fact, by doubling the permeability from 125 to 250, which is a practical value when impedances of less than
200 Ω are involved, the number of turns can be reduced by 30% while maintaining
the same low frequency response. This in turn increases the high frequency
response by 40% since it is inversely proportional to the number of turns. Experiments by Sevick have shown that toroids with outside diameters between 1.5 and 2

in can be used in most Amateur Radio applications and still handle the full legal limit of power.

When a rod is used as a transformer core, the calculations for the magnetizing inductance become complicated because of the effect of the high reluctance air path external to the core. As will be explained shortly, the magnetizing inductance is independent of the rod's permeability. Experimentally, it is on the order of one-half the value of a toroid with a permeability of 125.

With the following definition for available power:

$$P_{available} = \frac{V_g^2}{4R_g} \tag{4-5}$$

the equation for the low frequency performance of Figure 4-5 can be written as

$$\frac{P_{available}}{P_{out}} = \frac{R_g^2 + 4X_M^2}{4X_M^2} \tag{4-6}$$

where

$$X_M = 2\pi f L_M \tag{4-7}$$

It is apparent from equation (4-6) that the output power approaches the available power when X_M is greater than R_g. Even a factor of five produces a loss of only 1%; the smaller the value of R_g, the smaller the requirement on X_M.

With power transmission line transformers, instead of the well-known 3 dB loss for the upper and lower cutoff frequencies a more practical figure is 0.45 dB. This represents a loss of about 10% and is equivalent to a standing wave ratio (SWR) of 2:1 when dealing with antennas.

Assuming a loss of 10% at the low frequency end, the reactance of the primary winding according to equation (4-6) is

$$X_M = 3R_g/2 \tag{4-8}$$

Solving for the number of primary turns using equations (4-4), (4-7), and (4-8), we get

$$N_p \cong \sqrt{\frac{2R_g \times 10^7}{f\mu_0(A_e/l_e)}} \tag{4-9}$$

An approximation is used because some numbers are rounded off and because of small variations in the permeability (μ_0). Experimental results have agreed to within 10 to 20% of the values predicted by equations (4-6) and (4-9).

4.3   High Frequency Characterization

Ruthroff used a single basic building block for his 1:4 balun and unun transformers. By connecting the bifilar windings in two different ways, each summing a direct voltage and a delayed voltage via the transmission line, he was able to obtain these

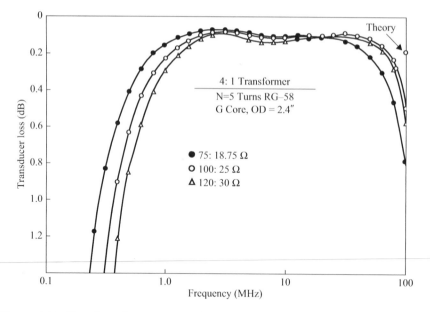

Figure 4-7   The plots show insertion loss versus frequency at three different impedance levels for a Ruthroff 4:1 transformer using 50 Ω coax.

two broadband transformers. In most cases these transformers are capable of handling the maximum power level specified for amateur radio use in the high frequency band and beyond. Because of the phase difference in Ruthroff's method of combining voltages, another factor came into the determination of the high frequency response—the length of the transmission line. In fact, at the frequency at which the electrical length of the transmission line becomes a half-wavelength, the high frequency response of the transformer is zero. Therefore, short coiled transmission lines, with sufficient inductance to satisfy the low frequency limit, are required to obtain wideband responses from his transformers.

Pitzalis and Couse [1] showed that experiment and theory agree quite well for a 4:1 transformer at the 100:25 Ω level when using a coiled 50 Ω cable. Sevick repeated this experiment, and the data are presented in Figure 4-7. The results show that the 50 Ω cable is optimum for a 100:25 Ω transformer.

Reference

[1] Pitzalis, O., and T. P. Couse, "Practical Design Information for Broadband Transmission Line Transformers," *Proceedings of the IEEE*, Apr. 1968, pp. 738–739.

Chapter 5

Transmission Line Construction

5.1 Introduction

We saw in Chapters 3 and 4 that the optimum characteristic impedance for a step-up or step-down transformer is the geometric mean between the source impedance and the load impedance. This requires us to use impedances that, in general, are nonstandard or hard to find commercially.

5.2 Commercial Transmission Lines

Commercial transmission line manufacturers produce readily available coaxial cables in 50, 70, 75, and 93 Ω impedances. Commercial cable is also available in impedances below 50 Ω, but the only source I could find was Communications Concepts, which has small cables available in 10.7, 17, 18.6, 22, 25, and 26 Ω. These are all useful for input transformers for amplifiers or low power ununs or baluns up to perhaps 200 W.

5.3 Custom Coaxial Transmission Lines

It is possible to create your own custom coaxial transmission line using commercial coax, magnet wire, polyimide tape, and braided shield. Sevick performed experiments on various methods of custom cable construction. The cables used inner conductors of different sizes, insulations of different thicknesses, and outer braids that were taped or bare. The object was to find various characteristic impedances from 10 to 35 Ω. Table 5-1 shows the results when the outer braid is tightly taped. (3M no. 92 tape is a good choice.) These are the values when wound around a toroid. When measured as straight sections of coax, the results are about 5% higher. Without taping the outer braid, and when the cables are wound around a toroid, the results are about 25% higher than shown in Table 5-1. The spacing between the inner conductor and the outer braid is not symmetrical under these conditions because of the tight bend radius. For example, a no. 14 wire with two pieces of no. 92 tape (resulting in four layers of this 2.8 mil thick tape when wound edgewise) becomes 22 Ω (instead of 18.5 Ω) without the outer braid being taped. It is likely that interaction between the core and the fields inside the coax is present due to imperfect shielding of the braid.

Table 5-1   Characteristic Impedance of Coaxial Cables Using Various
Combinations of Inner Conductors (Formex Coated), Insulators,
and Outer Braids (One Layer of 92 Tape)

Insulators and Outer Braids	No. 12 Wire (Ω)	No. 14 Wire (Ω)	No. 16 Wire (Ω)
2 layers of no. 92 tape, RG-122/U braid	12.5	14	19.5
4 layers of no. 92 tape, RG-122/U braid	15	18.5	22.5
6 layers of no. 92 tape, RG-122/U braid	17.5	21	26
2 layers of no. 92 tape and 2 layers of no. 27 (glass tape), RG-122/U braid	21	23.5	31
2 layers of no. 92 tape and 3 layers of no. 27 (glass tape), RG-122/U braid	23	26	35
2 layers of no. 92 tape and 5 layers of no. 27 (glass tape), RG-58/U braid	31	35	—

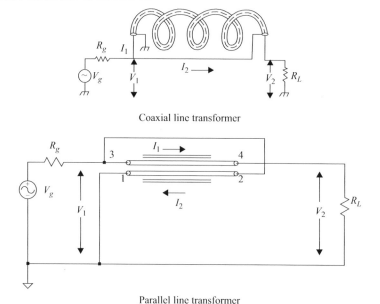

Coaxial line transformer

Parallel line transformer

Figure 5-1   Schematics show a comparison of a coaxial transformer and a parallel
line transformer.

The schematic in Figure 5-1 shows an example of a 4:1 low impedance coaxial cable transformer. The inner conductor used no. 12 wire and two pieces of no. 92 tape (wound edgewise, giving about four layers). The outer braid, which was bare, came from RG-58 coax. The transmission loss measurements in Figure 5-2 illustrate the optimum impedance level to be 50:12.5 Ω, which predicts a characteristic impedance of 25 Ω. From Table 5-1, a tightly taped braid with similar inner

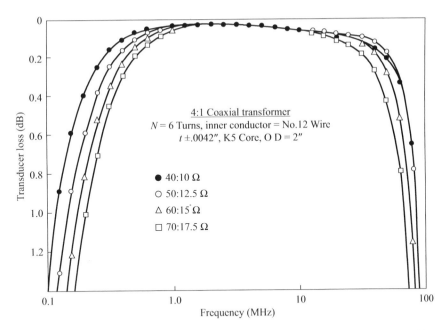

Figure 5-2 *Experimental results are shown for a low impedance coaxial cable transformer as a function of impedance levels. $Z_0 = 22$ Ω.*

conductor and insulation yields a value of 15 Ω. For a bare braid coax, it would increase to about 18.75 Ω (25% higher). Since a larger braid (RG-58 instead of RG-22) was used, and therefore more spacing occurred between the inner and outer conductors, the characteristic impedance rose to 22 Ω. As in most cases with low impedance coaxial cables, a value of 10% below the theoretical predictions for the optimum Z_0 was found to be common.

5.4 Custom Parallel Transmission Lines

Two wire transmission lines are generally constructed in one of three methods: two wires held parallel to each other; two wires twisted together; or two flat strips with the wide dimension one above the other.

Two wire parallel transmission lines are ideal for high impedance applications. Historically, high impedance lines in 300, 450, and 600 Ω impedances have been used for antenna feeders since the early days of radio. The equation for the characteristic impedance for an open wire line is

$$Z_0 = \left(\frac{1}{\sqrt{\varepsilon}}\right) 120 \cosh^{-1}\frac{D}{d}$$

$$Z_0 = \left(\frac{1}{\sqrt{\varepsilon}}\right) 276 \log_{10}\left(\frac{D}{d}\right) \quad \text{for } D \gg d$$

(5-1)

where ε is the dielectric constant of the material between the two conductors. Sevick referred to these lines as twin lead. Figure 5-3 shows the meaning of D and d.

Sevick observed that calculations for the characteristic impedance using a theoretical equation like equation (5-1) do not produce accurate results in the real world because of the uncertainties in spacing between the conductors, in the effects of the dielectrics, contributions of the magnetic core, and the proximity effect of neighboring turns. Sevick created 22 different configurations of wire size, spacing, and dielectric. These lines were wound around a toroid with significant spacing between the windings to minimize coupling between adjacent turns. The results are presented in Figure 5-4. The smallest spaces were obtained using 1.5 mil thin coated Formex wire. Various combinations of 1.5 mil thin coated wire and 3.0 mil thick coated Formex wire yielded spaces of 3.0, 4.5, and 6.0 mil. He then added no. 92 polyimide tape wrapped around the wires to create the intermediate spaces and then Teflon (DuPont trade name for PTFE) tubing to create additional spacing.

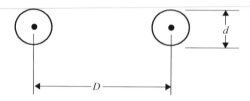

Figure 5-3   Illustration shows how to measure D and d from equation (5-1).

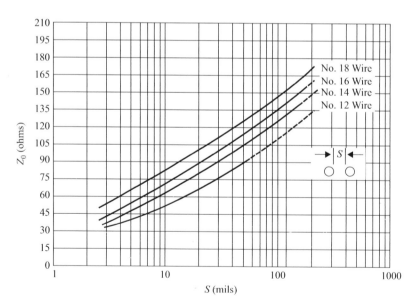

Figure 5-4   Curves show characteristic impedance, Z_0, versus wire diameter and spacing (S) for parallel transmission lines.

Finally, he created widely spaced lines by spacing the wires apart using narrow strips of no. 27 glass tape (7.5 mil thickness) wrapped around the wires at large intervals (similar in concept to using plastic spreaders to create 600 Ω open wire line). Notice that the impedance obtained using no. 12 wire is not linear at the smaller spacing. This is attributed to the difficulty in maintaining truly straight conductors using such large wire. Figure 5-3 also shows that characteristic impedance exceeding 150 Ω are difficult to obtain because of the rather large spacing required between the wires, the effect of neighboring turns, and contributions of the ferrite to both permeability and permittivity. As an example, 300 Ω TV ribbon was wound on a 2.4 in OD toroid, and measurements revealed a characteristic impedance of only 200 Ω.

A variation of parallel wire lines is the twisted pair. In this construction, the wires are twisted together with a moderate number of twists (generally 3 to 10) per inch. Twisting results in lower characteristic impedance because of increased capacitance and inductance per inch.

Winding a twin lead line on a ferrite core includes the permeability and permittivity of the core in the field contributions to operation of the line. The fields inside the core will create a slower field propagating along the line than that propagating in the air above the line. This is exactly analogous to the operation of the fields surrounding a microstrip line. The result is very similar to closely coupling two microstrip lines. Twisted pair line operates differently when placed on the core. The twist periodically alternates the direction of both the magnetic and electric fringing fields around the conductors. The net effect is that the TEM fields of twisted pair are almost entirely contained between the conductors until the size of the twist becomes an appreciable portion of a wavelength at the frequency of operation. The twisting prevents a propagating field from occurring within the ferrite of the core. Sevick observed this in some of his experiments but did not attribute it to the field in the ferrite.

Sevick used another parallel line construction he erroneously referred to as stripline. Figure 5-5 compares the constructions of what is called stripline by RF professionals, microstrip line, and the flat line Sevick used. I will refer to Sevick's

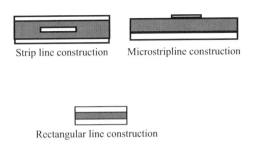

Strip line construction   Microstripline construction

Rectangular line construction

Figure 5-5   Cross sections show a comparison of strip line, microstrip line, and rectangular line construction. All three types use flat (rectangular) conductors in combination with a dielectric (gray material) to create a transmission line.

design as rectangular line since there is no industry standard definition for such lines. Stripline is manufactured so that the fringing fields between the center conductor and the outer conductors are completely contained within the dielectric (for practical purposes). Microstrip, likewise, is built so that the fringing fields are mostly (but definitely not completely) contained within the dielectric between the two conductors. You can see that Sevick's lines are very similar to microstrip with the exception that the bottom conductor of the line is not larger than the top. The fringing fields at the edges of the line have an appreciable contribution to operation as a transmission line. The end result is that the permeability and permittivity of the ferrite combine to lower the characteristic impedance below that of the line in free space.

Lossless transmission lines have characteristic impedance defined by

$$Z_0 = \sqrt{L/C} \tag{5-2}$$

where

L = the distributed inductance
C = the distributed capacitance

Z_0 can be lowered by increasing C, by lowering L, or by a combination of both. One method to obtain a low value of Z_0 is to use closely spaced flat conductors to produce a rectangular line. If the width of the flat conductor is much larger than the spacing between the conductors, the value of the characteristic impedance is

$$Z_0 = 377 \frac{t}{\sqrt{\varepsilon}W} \tag{5-3}$$

where

t = the spacing between the conductors
W = the width
ε = the dielectric constant of the insulation

To experimentally determine the optimum impedance levels for various widths of rectangular line, Sevick constructed four transformers and measured them at various impedance levels that bracketed the optimum level. The rectangular line transformers were constructed with widths of 1/8, 3/16, 1/4, and 3/8 in. Sevick did not describe the thickness of the copper used. The insulation was one layer of no. 92 tape. The cores were TDK K5 ferrite with an OD of 1 3/4 in. Figure 5-6 shows the loss versus frequency curves taken at their optimum levels, that is, impedance levels where the high frequency response is maximum. A rectangular line of about 7/64 in wide should be optimum for 50:12.5 Ω operation.

K&S Metals sells various metal products through hobby stores and other retailers. The longest rectangular line that you could make using K&S copper sheet is 12 in. K&S copper sheets are available in 6 in × 12 in sizes in 16 mil and 25 mil thickness, and purpose-built strips can be made using an ordinary paper shear. I found copper ribbon in 5, 10, and 20 mil thickness in widths of 1 or 1/2 in from Basic Copper at prices that are equivalent to buying 1/2 lb of magnet wire.

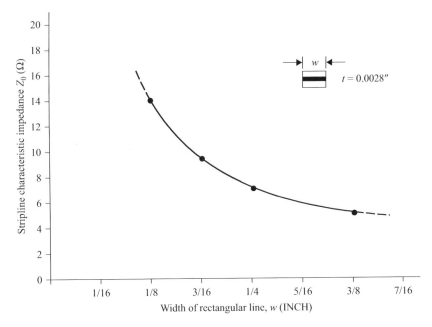

Figure 5-6   Plot shows characteristic impedance versus width for a rectangular line with 0.0028 in polyimide dielectric. The lines were wound on K5 cores and terminated in their characteristic impedance.

Table 5-2   Copper Wire to Copper Foil Equivalent Widths at DC

AWG Wire Size	Diameter (mil)	Area (mil²)	2.6 mil Tape Equivalent Width (in)	3.5 mil Tape Equivalent Width (in)	4 mil Tape Equivalent Width (in)	10 mil Tape Equivalent Width (in)	20 mil Tape Equivalent Width (in)
12	80.8	5128	1.972	1.465	1.282	0.513	0.256
14	64.1	3227	1.241	0.922	0.807	0.323	0.161
16	50.8	2027	0.780	0.579	0.507	0.203	0.101
18	40.3	1276	0.491	0.364	0.319	0.128	0.064
20	32	804	0.309	0.230	0.201	0.080	0.040

These ribbons will need a heavy duty shear or tin snips to cut to smaller widths. 3M manufactures copper foil tape in thicknesses of 2.6 mil (1181), 3.5 mil (1182), and 4.0 mil (1245) for electromagnetic compatibility (EMC) shielding purposes. They are available in widths of 1/8, 1/4, 3/8, 1/2, 5/8, and 3/4 in and vary by product type. Table 5-2 shows various American wire gauge (AWG) sizes and the equivalent widths of tape. Since the total area will be much larger when skin depth is taken into account, you can probably use the equivalent of two sizes smaller tape for equivalent current capacity at RF. For example, 1/2 in 4 mil tape should be roughly equivalent to AWG no. 12 wire.

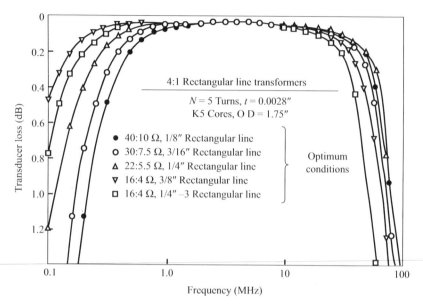

Figure 5-7   *Curves show transformer loss measurements using rectangular line for optimum characteristic impedance. The last line is a stack of alternating 1/4 in copper strips and polyimide tape with the outer strips connected in parallel at both ends in a fashion similar to stripline.*

The four rectangular line transformers were disconnected as transformers and measured *in situ* for their characteristic impedances as simple transmission lines. A plot of the results is shown in Figure 5-7. When relating the measured values in Figure 5-7 to the optimum impedance levels (at the same width) in Figure 5-6, the optimum characteristic impedance is found to be about 30% lower than Ruthroff predicted. For example, with 1/8 in rectangular line at the 40:10 Ω impedance level (which is optimum), his theory predicted Z_0 to be 20 Ω. Experimentally, it was found to be 14 Ω. This percentage difference is the largest obtained from any form of transmission line. Differences from theory with low impedance coaxial cables are usually less than 10%. With wire transmission lines, the differences are negligible. These differences are probably because of two things: (1) parasitic elements not included in the theory; and (2) end effects with short rectangular lines.

5.5   Closely Wound Twin Lead

The section describing twin lead on toroids assumes that the twin lead is widely spaced by at least the width of the line to avoid coupling between the opposite conductors as it wraps around the core. Another option is to wind the twin lead so that the conductors are immediately adjacent, which Sevick referred to as closely wound lines. Unless the core is very large, it is very difficult to make the wires immediately adjacent except on the inside of the core. This limits the mutual coupling between opposite conductors. The situation is different if a rod is used as the magnetic element for the transformer. In that case, each turn is in intimate contact with the previous and subsequent turn. The phase shift around the rod for each turn causes a slight opposition to the field propagating transversely along the line. The phase shift increases with larger diameter rods. Electromagnetic modeling should indicate that the fields traverse along the rod as well as around the rod. The closely wound lines on rods are capable of producing lower characteristic impedance with smaller wires. Rod transformers require many more turns compared with a toroid to have sufficient magnetizing inductance for the low frequency end. This reduces the high frequency end of the response in a Ruthroff transformer.

An early assumption that a tightly wound toroidal transformer had the same characteristic impedance of a rod transformer led to the conclusion that the optimum characteristic impedance for best high frequency response should be considerably lower than predicted. Actual measurements of the characteristic impedances on toroids (before the transformers were connected as ununs) showed the values to be greater by about 30% over the rods. This is attributed to the difficulty of achieving as tight a winding as possible on a toroid and the different effect of the fringing field on toroidal cores. Transmission loss measurements on toroids again showed that the optimum impedance level for maximum high frequency response was well predicted by Ruthroff.

Sevick found that tightly wound rod transformers did agree quite well with Ruthroff's theory. A characteristic impedance of 25 Ω, using no. 14 wire, yielded the best high frequency response at about the 12.5:50 Ω level. A 5/8 in diameter rod had the same high frequency response as a toroidal transformer! To understand this discrepancy, the characteristic impedance was measured on a five-turn transmission line of no. 14 wire, tightly wound on a 5/8 in diameter rod. The impedance was found to be 26 Ω (4 Ω higher than expected). In addition, connected as a 1:4 unun, the best high frequency response occurred at the 16.5:66 Ω level. Theoretically, the highest frequency response should be at the 13:52 Ω level. Other experiments with rod transformers, when varying the diameter of the rods and the number of turns to maintain the same low frequency response, showed that the characteristic impedance is directly related to the diameter and that the high frequency response is directly related to the ratio of the length of the coil to its diameter. In other words, the smallest diameter rods gave both the lowest characteristic impedance and the highest frequency response at the 12.5:50 Ω impedance level. Figure 5-8 shows the results of the characteristic impedance measured on rods of varying diameters with closely wound transmission lines of no. 14 wire.

5.6   Three Conductor Lines

Sevick found that low impedance lines are relatively easy when a rod is used for the transformer but not when a toroid is used. We know that we can create a low

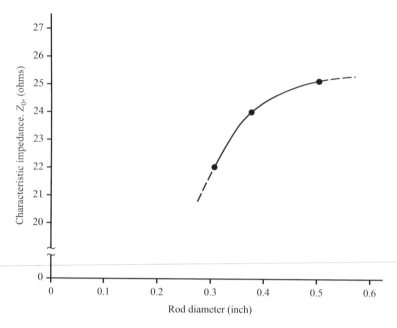

Figure 5-8   Plot shows characteristic impedance versus rod diameter for closely wound twin lead using no. 14 wire.

impedance transmission line by connecting two lines in parallel. If we instead create a parallel line with three conductors side by side, we can create the equivalent of two lines in parallel. With a two-wire line, the fields are contained between the two conductors. Adding the third line and connecting the two outer lines in parallel now gives an equal set of fields between the center conductor and the new outer conductor. The effect is to create a new line with one-half the impedance of an equivalent two-wire line. Figure 5-9 shows the schematic for this transformer.

Figure 5-10 shows the experimental data for a 4:1 transformer with five trifilar turns of no. 14 wire wound on a G core. The transformer is now optimized at the 75:18.75 Ω level. Even the 50:12.5 Ω level performance is much better than that of the closely wound transformer in Figure 5-11. Adding more wires in parallel improves the low impedance level response even further. This multiwire configuration approaches the performance limit of coaxial cable.

The second method uses the third wire in a floating connection (Figure 5-12). The third wire modifies the fields and lowers the impedance of the combined structure. This result was found experimentally while observing the performance of 9:1 transformers (input connected to terminal 3 and output to terminal 6). The experimental results are shown in Figure 5-13 to demonstrate the dramatic impact of the floating third wire. The results for the 4:1 connection at the 50:12.5 Ω level is extraordinarily good for a wire transformer with a toroidal core.

The last plot in Figure 5-6 shows the results of another floating third-wire transformer. The trifilar rectangular line uses 1/4 in wide strips, with the third strip

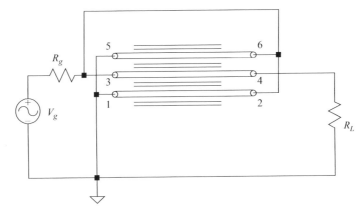

Figure 5-9   The schematic shows a three-wire 4:1 Ruthroff transformer. This configuration lowers the characteristic impedance because the center conductor carries twice the current of each of the outer wires.

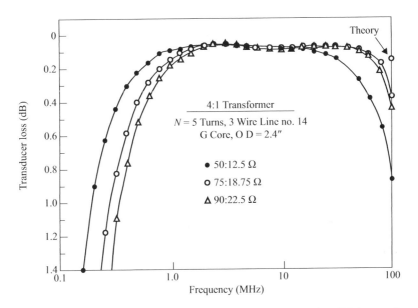

Figure 5-10   Curves show the loss versus frequency for a Ruthroff 4:1 transformer using a three-wire line with the outer wires connected in parallel.

floating. It compares favorably to the bifilar rectangular line using 3/8 in strips. By using a floating strip, the optimum condition for 1/4 in rectangular line was lowered from 22:5.5 to 16:4 Ω. It is interesting to note that the 3/8 in two-conductor rectangular line and the 1/4 in three-conductor rectangular line have their best high frequency performance at the low impedance level of 16:4 Ω. And finally, when designing transformers for low impedance applications, the currents can be very large

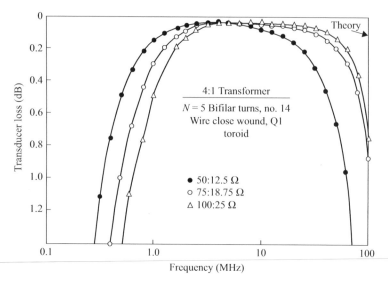

Figure 5-11   The loss of a transformer similar to that in Figure 5-10 is shown but with bifilar construction instead of trifilar.

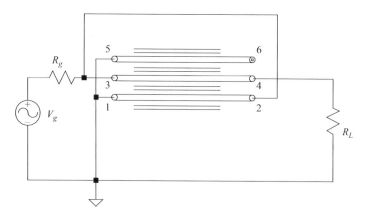

Figure 5-12   The schematic of a trifilar wound transformer is shown with the third wire left open at the load.

at high power levels. Therefore, wide parallel lines and coaxial cables have definite advantages over wire lines, since the currents are evenly distributed on the conductors. With wire transformers, the currents are crowded between adjacent turns.

5.7   Custom Multiconductor Transmission Lines

Sevick explored configurations consisting of four and five wires placed in parallel around rods or toroids. The interactions between the outer two wires in those

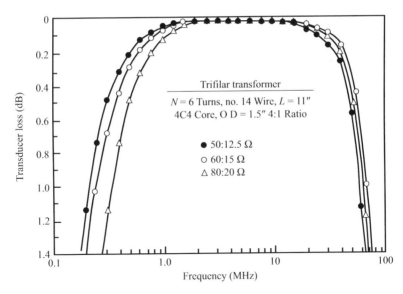

Figure 5-13   *Experimental results are shown for a trifilar transformer designed for 4:1 operation. The transformer shows the excellent frequency response of a wire transformer at the 50:12.5 Ω level because of the floating third wire.*

configurations are intuitively quite different from the interactions of any two of the wires that are immediately adjacent. It requires very expensive electromagnetic modeling software to determine the interactions between all of the wires in a four- or five-wire configuration. Sevick performed empirical studies to determine the possibilities. His experiments showed the interactions provided for good low impedance performance in noninteger transformer ratios. His results are presented in later chapters.

5.8   Comparison of Twisted Pair and Twin Lead

The discrepancies noted in section 5.3 regarding the optimum characteristic impedance and the expected high frequency performance prompted Sevick to perform further experimental investigations. He compared twisted pairs with other types of windings: (1) no. 16 twisted pair (five turns per inch) with a characteristic impedance of 40 Ω; (2) tightly wound no. 16 wire with a characteristic impedance of 35 Ω; and (3) a pair of no. 16 wires in twin lead configuration held closely together by insulating tape and with a characteristic impedance of 50 Ω. Figure 5-14 shows the experimental results as a function of impedance levels for these 4:1 transformers.

On balance, the results in Figure 5-14 favor the simple twin-lead winding. Surprisingly, the twisted pair was no better than the twin-lead winding at the lowest

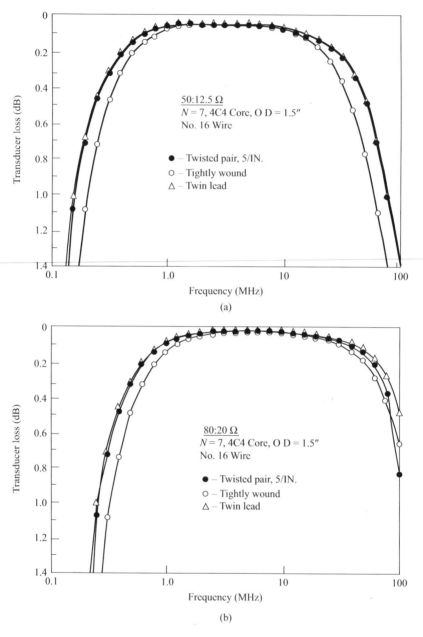

Figure 5-14   Sets of curves show a comparison of performance versus frequency of twisted pair, tightly wound, and twin-lead transformers at four different impedance levels: (a) 50:12.5 Ω. (b) 80:20 Ω. (c) 100:25 Ω. (d) 120:30 Ω.

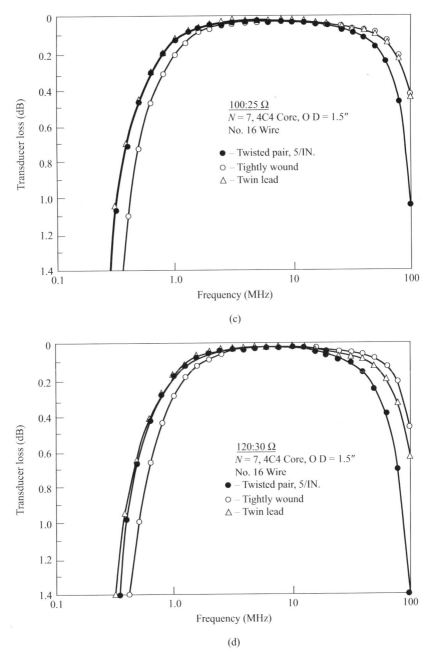

(c)

(d)

Figure 5-14   (Continued)

impedance level of 50:12.5 Ω, and the tight winding was the best at the highest level of 120:30 Ω. The poorer low frequency performance of the tightly wound transformer is attributable to the difference in the permeability of its core. This experiment points out not only the parasitic effects not considered in the theory but also the questionable use of twisted pairs for obtaining better low impedance performance. However, Ruthroff also used very fine wires (nos. 37 and 38 Formex) and extremely small cores (0.175 to 0.25 in OD). Twisting was necessary to keep the wires closely coupled. Another limitation is encountered in the high impedance use of these transformers, especially where high efficiency is important. As was shown in Chapter 2, most ferrites exhibit greater loss at higher impedance levels. Sevick attributed the loss to dielectric loss, but it is also due to hysteresis loss at the higher flux density because of the higher number of volts per turn at higher impedances.

5.9 Long Transmission Lines

Blocker presented an analysis that predicted a good match over a broad frequency band for a wide range of impedance levels [1]. His analysis was applicable only for impedance ratios equal to or greater than 4:1. By using Ruthroff's equations for output power, available power, and characteristic impedance for maximum high frequency response, he arrived at

$$P_{out} = 4(R_g/R_L)(1 + \cos(\beta l))^2$$

$$P_{available} = \left[2(R_g/R_L)(1 + \cos \beta l) + 1\right]^2 - \sin^2 \beta l$$

(5-4)

A perfect match, $P_{out} = P_{available}$, occurs in equation (5-4) whenever the electrical length of the line βl and the ratio R_L/R_g are related by

$$\sec \beta l = R_L/2R_g - 1$$

(5-5)

This condition can be satisfied for any arbitrary value of R_L/R_g that is equal to or greater than 4. At a given frequency, the higher transformation ratios require longer lines for a perfect match. Table 5-3 gives the values of $P_{out}/P_{available}$ as a function of line length for values of R_L/R_g ranging from one to infinity. The table assumes that the characteristic impedance of the line is always adjusted to the relationship $Z_0 = \sqrt{R_g R_L}$. The table shows that the perfect matches for ratios greater the 4:1 require considerably longer transmission lines. For example, a 16:1 transformer requires a length of 0.227 λ. This is a very long length of transmission line, even at 30 MHz. It certainly is not realizable on a conventional core. However, transformers in the 100 MHz region can utilize this technique.

5.10 Variable Characteristic Impedance Lines

The use of tapered lines in wide band matching at very high frequencies is well known [2]. Roy showed analytically that it is possible to achieve matching at the

Table 5-3 $P_{out}/P_{available}$ *as a Function of Line Length ($\beta l/2\pi$) and the Ratio of Load to Source Resistance (R_L/R_g)*

R_L/R_g	1.0	2.0	3.0	4.0	5.0	6.0	9.0	16.0	25.0	∞
$1/2\pi$										
0.000	0.640	0.889	0.980	1.000	0.988	0.960	0.852	0.640	0.476	0.000
0.134	0.609	0.847	0.949	0.990	1.000	0.993	0.934	0.768	0.610	0.000
0.167	0.590	0.818	0.923	0.973	0.994	1.000	0.973	0.852	0.714	0.000
0.204	0.559	0.768	0.871	0.928	0.962	0.982	1.000	0.964	0.888	0.000
0.227	0.532	0.723	0.820	0.876	0.914	0.939	0.979	1.000	0.988	0.000
0.236	0.520	0.703	0.795	0.850	0.887	0.912	0.956	0.992	1.000	0.000
0.250	0.500	0.667	0.750	0.800	0.833	0.857	0.900	0.941	0.962	1.000
0.333	0.308	0.333	0.324	0.308	0.290	0.273	0.229	0.165	0.121	0.000
0.500	0.000	0.000	0.000	0.000	0.000	0.000	0.000	0.000	0.000	0.000

lower frequencies of a 4:1 transformer over a wider band than is possible by using a uniform line. He considered the exponential line

$$Z_c(x) = Z_0 e^{(2\tau x/l)} \tag{5-6}$$

where

Z_0 = the characteristic impedance at $x = 0$
τ = the taper parameter (near unity in most cases)
l = the length of the transmission line

Since tapering can best be achieved in microstrip form, this interesting technique is best left to professionals with sophisticated resources. To date, no experimental data have been made available.

Irish approached the wideband matching problem by using a transmission line that varied as a step function along the length of the line [3]. The analysis he presented was for a line of two sections with differing characteristic impedances. The dimensions of the bifilar winding were changed midway along its length. Analytically he showed an extension by about 30% of the useful bandwidth was possible. Again, no experimental data were given.

References

[1] Blocker, W., "The Behavior of the Wideband Transmission Line Transformer for Nonoptimum Line Impedance," *Proceedings of the IEEE*, Vol. 65, 1978, pp. 518–519.

[2] Collins, R. E., *Foundations for Microwave Engineering*, New York, McGraw Hill, 1966, Chap. 5.

[3] Irish, R. T., "Method of Bandwidth Extension for the Ruthroff Transformer," *Electronic Letters*, Vol. 15, Nov. 22, 1979, pp. 790–791.

Chapter 6

1:4 Unun Transformer Designs

6.1   Introduction

In the literature, the 1:4 unun transformer has been analyzed the most, with credit given in large measure to Ruthroff. It finds extensive use in solid-state circuits and in many antenna applications when matching ground-fed antennas such as shortened verticals, vertical beams, slopers, and inverted L antennas, where impedances of 12 to 13 Ω have to be matched to 50 Ω coax cable. Even very short vertical antennas, which are used for mobile operation, can approach impedances of 12 Ω because of losses in the loading coils and the less-than-perfect ground systems.

These 1:4 transformers can be designed basically in two ways: (1) the Ruthroff method, which uses a single, coiled transmission line and a feedback (or bootstrap) connection to sum two voltages; and (2) the Guanella method, which sums two voltages using two or more coiled transmission lines (and thus equal delays) in a parallel–series connection. The Guanella transformer, which is basically a balun, requires extra isolation as an unun.

There are several rules of thumb concerning design of ununs:

1.   They are sensitive to impedance levels; they are designed for either high or low impedance applications.
2.   The high impedance design (which can mean impedances of only 300 to 400 Ω) is by far the more difficult to construct just because of size limitations; a 50:200 Ω unun requires twice the number of turns of a 12.5:50 Ω unun for the same low frequency response.
3.   For high efficiency operation, low permeability ferrites (100 to 300) are to be used.
4.   Rod transformers, even in the Ruthroff method, can find extensive use in the 1.5 to 30 MHz range when used in low impedance applications, but the rod transformer generally requires twice the number of turns of a toroidal transformer.

The purpose of this chapter is to provide examples of 1:4 unun designs. In many cases, the ferrite rods used in these examples came from AM loop stick antennas and filament chokes from class-B linear amplifiers. They can be found in flea markets and amateur radio junk boxes and are also readily available commercially from Elna Magnetics, Amidon, and RF Parts. These rods generally have a permeability of 125

and are ideal for this use. The toroids in these designs came from many manufacturers. Generally, all of the nickel-zinc (NiZn) ferrite materials (with the proper
permeability) from the various suppliers have been found to be acceptable. The
choice should be made on the basis of availability and price. The one exception is
4C4 material from Ferroxcube. It is the hardiest of all the ferrites tested by Sevick
and is recommended in applications where damage can come from very high impedances, such as those experienced in antenna tuners. In this case, the magnetizing
inductance can become a part of the resonant circuit, thus creating a damaging high
flux density in the core. It is also important to remember that references in this
chapter to material 52 apply to Fair-Rite ferrite material and *not* Micrometals powdered iron material 52. It is easy to tell which type of 52 material you have: powdered
iron is color coded as green with blue on one face; ferrite is always a flat gray color.

6.2   Schematics and Pictorials

Figure 6-1 shows the schematics and pictorials for the 1:4 unun transformer using
the single transmission line in the Ruthroff design. In the figure the generators are
at the low impedance side in a step-up operation. Since these transformers are
linear and bilateral, the generators could just as well have been placed on the high

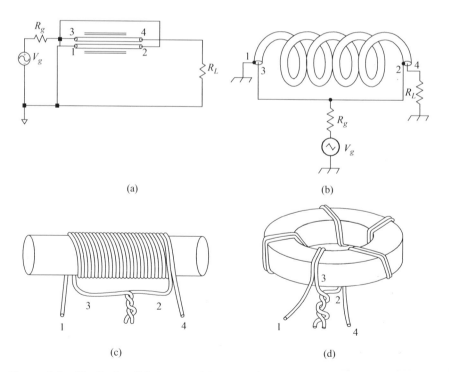

(a)   (b)

(c)   (d)

*Figure 6-1   The Ruthroff 1:4 unun: (a) Wire schematic. (b) Coax cable schematic.
(c) Rod construction pictorial. (d) Toroid construction pictorial.*

side (on the right) in a step-down operation. The main consideration is that the characteristic impedance, Z_0, of the coiled transmission line is one-half the value of the high impedance side and twice the value of the low impedance side. A coax cable representation is not included but can be readily understood from Figure 6-1c and Figure 6-1d. Many photographs shown throughout the text will also help in visualizing the various windings and connections.

Two versions of the 1:4 unun transformer using the Guanella method of adding two voltages of equal delays are illustrated in Figure 6-2. Since the Guanella transformer is basically a balun design, extra isolation has to be considered when operating it as an unun transformer. Figure 6-2a shows such an arrangement utilizing a 1:1 balun in series with a 1:4 balun. With sufficient isolation from the 1:1 balun, the input of the 1:1 balun and the output of the 1:4 balun can both he grounded, resulting in very broadband unun operation.

The second version (Figure 6-2b) basically uses a single core with only the top transmission line wound on it. When the characteristic impedance of each transmission line has the optimum value, $Z_0 = R_1/2$, the bottom transmission line has no potential gradient from input to output and therefore requires no magnetic core. The core gives only mechanical support. When the characteristic impedance departs considerably from the optimum value, then winding the bottom transmission line on a magnetic core would improve the low frequency response. The major requirement

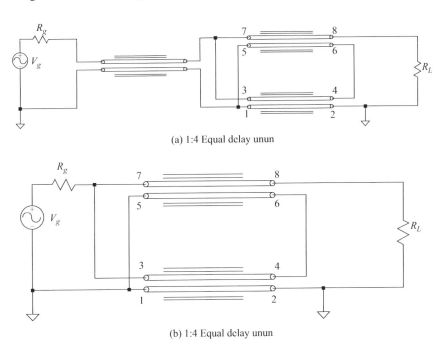

(a) 1:4 Equal delay unun

(b) 1:4 Equal delay unun

Figure 6-2   *This schematic shows two versions of the Guanella 1:4 unun: (a) A 1:1 balun back to back with a 1:4 balun. (b) Windings on two separate cores (in which the bottom wire acts only as a delay line).*

here is that the reactance of a single winding be much greater than the low impedance side of the transformer. If the reactances of windings 5–6 and 7–8 in Figure 6-2b have the same values as those of the Ruthroff unun in windings 1–2 and 3–4 in Figure 6-1a, then the two transformers will have the same low frequency responses. Since Guanella's transformer in Figure 6-2b adds in-phase voltages (windings 1–2 and 3–4 acting only as a delay line), its high frequency response will be considerably higher than Ruthroff's in Figure 6-1a.

Even though the Guanella ununs are more complicated than that of the single transmission line approach of Ruthroff, measurements made on these transformers, using the simple test equipment in Chapter 12, show much less phase shift at the higher frequencies and hence greater high frequency response. They also lend themselves more readily to combination balun/unun use.

6.3   12.5:50 Ω Ununs

Figure 6-3 shows four Ruthroff designs (presented as diagrams in Figure 6-1A) using rods with diameters from 1/4 to 1/2 in. The cable connectors are all on the low impedance sides of the transformers. The length of the rods, which is not critical, varies from 2 to 4 in. The permeability of each is 125—the same as that of the ferrite in the AM loop stick antenna. The two transformers on the left, which use no. 14 wire, are capable of handling 1 kW continuous power. The two on the right, which use no. 16 wire, are capable of handling 200 W continuous power. A single

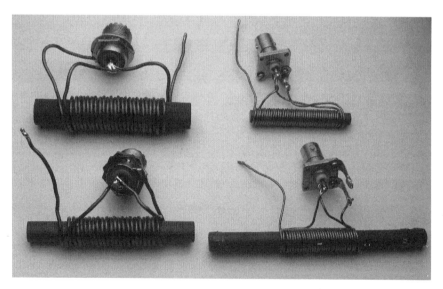

Figure 6-3   Four Ruthroff 1:4 ununs designed to match 12.5 to 50 Ω in the frequency range of 1.5 to 30 MHz. The two on the left are rated for 1 kW continuous power, and the two on the right are rated at 200 W continuous power.

Formvar (SF) coated wire was used successfully. Others, like Formex and PE (plain enamel), should find equal success. If a more conservative design is needed, then Pyre ML or H Imidez (which are the same) is recommended. The latter two have thick coatings of an aromatic polyimide (about 3 mil). Tightly wound transformers with these polyimide coatings, such as those in Figure 6-3, would have breakdowns similar to those of coax cables. Further, the differences in characteristic impedances of the coiled transmission lines due to the extra 1.5 mil thickness are negligible.

All four rod transformers in Figure 6-3 were optimized for operation at the 12.5:50 Ω level in the 1.5 to 30 MHz range and have the following parameters:

Upper left: 1/2 in diameter, 11 bifilar turns of no. 14 wire
Lower left: 3/8 in diameter, 14 bifilar turns of no. 14 wire
Upper right: 1/4 in diameter, 20 bifilar turns of no. 18 wire
Lower right: 3/8 in diameter, 16 bifilar turns of no. 18 wire

Toroidal transformers have the advantage of a closed magnetic path and thus require fewer turns compared with rod versions to attain the desired reactance of the coiled transmission line. The shorter winding length improves the high frequency response of Ruthroff 1:4 transformers. However, they have the disadvantage of not being able to achieve a characteristic impedance of 25 Ω with closely spaced bifilar turns, as can the rod transformers in Figure 6-3. Therefore, other types of transmission lines, such as low impedance coax cable, rectangular line, or floating third winding (Chapter 5), must be used in 1:4 transformers at the 12.5:50 Ω level and lower. Sevick chose not to include rectangular line transformers in his evaluation because he believed them to be too difficult for the average experimenter to obtain. As we will see in Chapter 13, rectangular line using 16 or 25 mil copper is easily constructed for short windings, and longer rectangular line can be created easily by splitting 4.0 mil EMC tape.

Even though low impedance coax cables are not readily available, they can easily be fabricated (Chapter 13). Coax cable and rectangular line have the advantage in handling higher power levels since the currents are not crowded between adjacent turns as in wire transformers. Further, two or more layers of 3M no. 92 tape are usually used as the dielectric for purpose built coax, thus yielding breakdowns comparable to RG-8/U cable. Many of these coax cable transformers are truly in the 5 kW range.

Four versions of 1:4 (12.5:50 Ω) unun coax cable transformers are presented in Figure 6-4. The cable connectors are on the low impedance sides of the transformers with the following parameters:

Lower left: 7 turns of coax using no. 16 wire for inner conductor, insulated with two layers of no. 92 tape. The outer braid is from RG-174/U cable. The toroid is 1.38 in OD 52 material ($\mu = 250$). Sevick originally created this transformer on a 1.25 in, K50 core ($\mu = 250$), but that material is no longer available.
Upper left: 8 turns of coax using no. 12 wire for inner conductor, insulated with two layers of no. 92 tape. The outer braid is from RG-58/U cable. The toroid is 2 in OD, 61 material ($\mu = 125$).

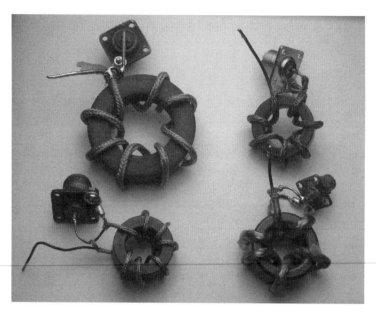

Figure 6-4   This photo shows four representative Ruthroff 1:4 toroidal transformers.

Upper right: 7 turns of coax using no. 14 wire for inner conductor, insulated with four layers of no. 92 tape. The outer braid is from RG-58/U cable. The toroid is 1 1/2 in OD, 4C4 material ($\mu = 125$).

Lower right: 6 turns of coax using no. 14 wire for inner conductor, insulated with six layers of no. 92 tape. The outer braid, which is tightly wrapped with no. 92 tape, is from RG-58/U cable. The toroid is 1 1/2 in OD, 52 material ($\mu = 250$).

The transformer in the lower left of Figure 6-4 has, by far, the widest bandwidth of the four, ranging from 1.5 to over 50 MHz. This is due to using a smaller toroid and one of the highest permeabilities ($\mu = 250$) while still offering high efficiency. This results in a very short transmission line—only 10 1/2 in long. A power test (see Chapter 12) showed that this small transformer was capable of handling 700 W without excessive temperature rise. A conservative rating would be 200 W continuous power. It is interesting to note that the two transformers on the right in Figure 6-4 use two different types of toroids: inner conductors and outer braids. The characteristic impedances of both coax cables are about 22 Ω even though the one on the lower right has two more layers of no. 92 tape. The difference is mainly caused by wrapping the outer braid with no. 92 tape. Wrapping the outer braid with tape reduces the spacing between the inner conductor and outer braid, which lowers the characteristic impedance by about 25%. Also, since the transformer in the lower right in Figure 6-4 uses a higher permeability toroid (250 compared with 125), one fewer turn is required to have about the same low frequency response. This in turn raises the high frequency response by a ratio of 7:6.

Figure 6-5 shows two versions of a floating-third-winding 1:4 unun using toroids. Without the third winding, the characteristic impedance would be on the order of 45 Ω and thus have the highest frequency response at the 22.5:90 Ω level. The third winding, which is left floating, reduces the characteristic impedance to about 30 Ω. This enables fairly good 12.5:50 Ω operation. Since the characteristic impedances are in excess of 25 Ω, only small toroids, resulting in the shortest possible lengths of transmission line, are recommended. The parameters of these two transformers are as follows:

Left: 7 trifilar turns of no. 14 wire on a 1.25 in OD, 52 toroid ($\mu = 250$).

Right: 9 trifilar turns of no. 16 wire on a 1.25 in OD, 4C4 toroid ($\mu = 125$).

The transformer on the left in Figure 6-5 is capable of handling 1 kW continuous power. The other can easily handle 200 W continuous power. The transformer on the left has higher permeability, so the length of transmission line is shorter. The shorter line results in a higher frequency response than the transformer on the right while still having approximately the same lower frequency response.

The preceding examples in this section used the Ruthroff 1:4 circuit of Figure 6-1a and were capable of covering the 1.5 to 30 MHz range at power levels common to amateur radio. These transformers used a single transmission line and, at the 12.5:50 Ω level, allowed for short enough transmission lines to satisfy both the low and high frequency requirements. The 1:4 Guanella transformers

Figure 6-5   This photo illustrates two examples of floating third-wire transformers.

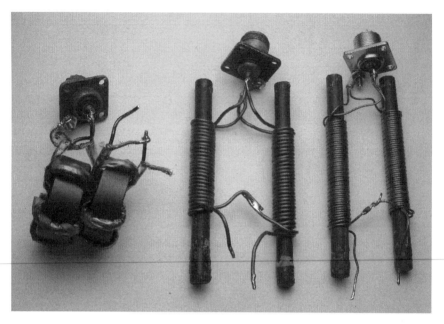

*Figure 6-6   These three transformers show examples of Guanella 1:4 unun
transformers.*

(Figure 6-2) have two transmission lines in a parallel–series connection, resulting in
the addition of two in phase voltages and giving much higher frequency capability.
The problem with the Guanella transformer is providing sufficient isolation when
operating as an unun transformer. This is especially true when both transmission
lines are wound on the same core. Although this yields the best low frequency
response when operating as a balun, the grounding of both the input and output
terminals makes it impractical as an unun.

As shown in Figure 6-2, this isolation can be obtained in two ways: (1) by
connecting a 1:1 balun back-to-back with a 1:4 Guanella balun which has both
transmission lines on the same core; and (2) by putting the two transmission lines on
separate cores. In the latter case, only one magnetic core is really needed, since the
bottom transmission line in Figure 6-2b has no longitudinal potential gradient.
Windings 5–6 and 7–8 determine only the low frequency response. If these windings
are the same as windings 1–2 and 3–4 in the Ruthroff 1:4 unun in Figure 6-1a, then
the Guanella 1:4 unun will have not only the same low frequency response but also a
much greater high frequency response. By using coax cable widely spaced on a core
(to minimize parasitic effects) or beaded, straight coax cable, the 1:4 unun transformers in Figure 6-2b should be capable of operating in the VHF and UHF bands.

Three examples of these very wideband 12.5:50 Ω Guanella unun transformers, using the schematics in Figure 6-2b, are illustrated in Figure 6-6, with the
following parameters:

Left: 6 turns of coax on each toroid with no. 14 wire for inner conductor, with six layers of no. 92 tape. The outer braids, which are tightly wrapped with no. 92 tape, are from RG-58/U cable. The toroids are 1 1/2 in OD, 52 material ($\mu = 250$).

Center: 14 1/2 bifilar turns on each rod of no. 14 wire. The rods are 3/8 in diameter and are no. 64 material ($\mu = 250$).

Right: 25 bifilar turns of no. 18 wire on 3/8 in diameter rods. The rods are no. 61 material ($\mu = 125$).

The transformers have continuous power ratings of 5 kW (left), 1 kW (center), and 100 W (right) and cable connectors on the high impedance side (50 Ω). Although not shown, in actual operation as a 1:4 unun, one of the low impedance output leads is grounded. The preference in the coax cable version (left) is to ground the outer braid (the strap connection).

6.4   25:100, 50:200, and 75:300 Ω Ununs

The 1:4 unun at the 25:100 Ω level presents an interesting case since it probably is the easiest one to design. The optimum characteristic impedance, Z_0, of 50 Ω is readily obtained from RG-58/U or wire with several layers of no. 92 tape. Further, toroidal transformers with a single transmission line (Ruthroff's design, Figure 6-1a) can still provide ample bandwidths. There are only a few applications at this impedance level. Most of the useful 1:4 ununs match 50 to 12.5 Ω, 50 to 200 Ω, or 75 to 300 Ω.

Figure 6-7 gives two Ruthroff ununs designed to operate at the 25:100 Ω impedance level. The cable connectors are at the low impedance sides of the transformers and have the following parameters:

Left: 9 bifilar turns of no. 14 wire on a 1 1/2 in OD, 4C4 toroid ($\mu = 125$). One wire is covered with two layers of no. 92 tape ($Z_0 = 50$ Ω).

Right: 14 bifilar turns of no. 16 wire on a 3/8 in diameter rod ($\mu = 125$). One wire is covered with 2 layers of no. 92 tape ($Z_0 = 54$ Ω).

The toroidal version in Figure 6-7 easily covers the 1.5 to 30 MHz range with margins. The 3/8 in diameter rod version can also cover the same frequency range, but with no margin at the low frequency end. If operation is predominantly in the 1.5 to 7.5 MHz range, then two or three extra bifilar turns should be added. This rod transformer is capable of handling 200 W continuous power.

A jump in impedance level to 50:200 or 75:300 Ω presents a much more formidable task in design. These higher impedance levels require reactance (to maintain low volts per turn) to be two and three times greater than the 25:100 Ω series and four and six times greater than the 12.5:50 Ω series. Therefore, the number of turns has to increase 40–70% over the 25:100 Ω series and 100–125% over the 12.5:50 Ω series. Since the high frequency response is inversely proportional to the length of the transmission line in the Ruthroff transformer (Figure 6-1a), it is very difficult to have high power handling and wideband response with his bootstrap technique.

Figure 6-7   *These transformers show examples of bifilar 1:4 Ruthroff transformers.*

One then has to resort to using a Guanella-based unun, which is not as sensitive to the length of the transmission line since in-phase voltages are summed. Also, since the Guanella transformer is basically a balun, it has to be adapted to unun operation (Figure 6-2).

Three versions of the Ruthroff transformer capable of working efficiently at the 50:200 Ω level are presented in Figure 6-8. They vary in wideband response and power handling capability:

Upper left: 12 bifilar turns of no. 16 wire on a 1 1/2 in OD, 52 toroid ($\mu = 250$). The sleeving is 15 mil wall plastic. The impedance ratio is fairly constant at 1:4 from 1.5 to 15 MHz and becomes greater than 1:5 at 30 MHz. A conservative power rating is 500 W continuous power. As before, a flatter response would be obtained with fewer turns (12 instead of 14).

Upper right: 11 bifilar turns of no. 22 hook up wire on a 5/8 in OD, 64 toroid ($\mu = 250$). The insulation is 12 mil wall plastic. At the optimum impedance level of 50:200 Ω, the frequency range is 1.5 (marginally) to over 65 MHz. The transformation ratio varies only from 1:4 to 1:5 over this wide frequency range. This is a result of the short transmission line—only 13 in long. The power capability is limited only by the melting of the thin wire, which could be 100 W or more.

Bottom: 13 bifilar turns of no. 16 wire on a 4 in diameter 61 rod ($\mu = 125$). The insulation is 15 mil wall Teflon tubing. At the optimum impedance level of

Figure 6-8   50:200 Ruthroff transformers require additional insulation.

50:200 Ω, the useful frequency range is 7 to 25 MHz. At 30 MHz, the impedance ratio rises to 1:5. The power rating is 500 W continuous power.

To attain a much higher frequency response at high power levels, the Guanella-based transformers shown in Figure 6-9 must be used:

Left: A 1:1 balun back-to-back with a 1:4 balun (Figure 6-2a). The 1:1 balun (the smaller toroid) has 11 bifilar turns of no. 14 wire on a 1 3/4 in OD 52 toroid ($\mu = 250$). One of the wires has two layers of no. 92 tape ($Z_0 = 50$ Ω). The 1:4 balun has two windings of 9 bifilar turns, in series aiding (wound in the same direction), of no. 14 wire on a 2.4 in OD, 64 toroid ($\mu = 250$). The wires are covered with 18 mil wall Teflon tubing. At the optimum impedance level of 55:220 Ω, the transformation ratio is flat from 1.5 to 50 MHz. A conservative power rating is 1 kW continuous power.

Right: Two toroids are used with: 15 bifilar turns of no. 16 wire; and 16 turns of no. 16 wire. The extra phase delay of the 16 turn winding is negligible. The windings, which are covered with 18 mil wall Teflon tubing, have a characteristic impedance of 125 Ω. The cores are 2.4 in OD, 64 material ($\mu = 250$). At the optimum impedance level of 62.5:250 Ω, a constant impedance ratio exists from 1.5 to well beyond 30 MHz. If no. 14 wire is used, this flat response would occur at the 50:200 Ω impedance level. A conservative power rating with either wire would be 1 kW continuous power.

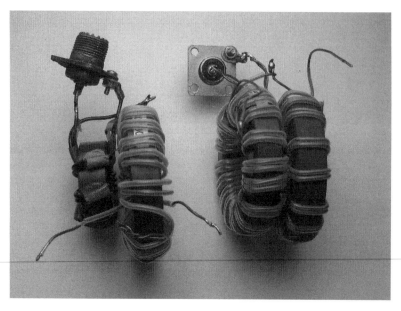

*Figure 6-9   This photo shows two Guanella unun transformers. The transformer
on the left uses the schematic of Figure 6-2a. Both transmission lines
of the 1:4 portion are wound on the same core with the wires in series
aiding. The transformer on the right is built using the schematic of
Figure 6-2b. Each transmission line is wound on its own core.*

In reviewing the differences, at the 50:200 Ω level, between the two types of
Guanella transformers shown in Figure 6-9, the advantage probably goes to the one
on the left, which uses two baluns back to back (Figure 6-2a). This permits a
smaller toroid to be used in the 1:1 balun. If the characteristic impedance of the 1:1
balun is 50 Ω, the same as that of the transmission line feeding this transformer,
then the 1:1 balun adds only about 2 ft to the feeder cable. It is essentially
transparent.

Chapter 7

Unun Transformer Designs with Impedance Ratios Less Than 1:4

7.1   Introduction

Little information is available on the characterization and practical design of transmission line transformers with impedance ratios less than 1:4. Investigations reported in the literature have treated the bifilar winding and its application to obtain impedance ratios of only 1:1, 1:4, 1:9, and 1:16. But many applications can be found for efficient, broadband transformers with impedance ratios of 1:1.5, 1:2, and 1:3. Some examples include the matching of 50 Ω coax cable to (a) vertical antennas, inverted L antennas, and slopers (all over good ground systems); (b) a junction point of two 50 Ω coax cables; (c) 75 Ω coax cable; and (d) shunt-fed towers performing as vertical antennas.

For many of his experiments described in this chapter, Sevick used K5 material, which has not been available for many years. Those wishing to experiment with transformers similar to Sevick's should consider using the easily accessible Fair-Rite material 52. K5 has permeability of 290, whereas 52 has permeability of 250. The difference will have some effect on the low frequency operation of the transformers, but the low frequency cutoff in most cases is below the HF bands.

Sevick's early work showed that the tapped Ruthroff 1:4 transformer can yield low impedance designs under certain conditions. Their designs and limitations are included in this chapter. His more recent work demonstrated that, by far, a better technique is to use higher order windings with the Ruthroff bootstrap method. This leads to further interesting applications such as broadband, multimatch transformers and compound arrangements, which opens up a new class of baluns capable of matching 50 Ω cable directly to the low impedance, balanced inputs of Yagi antennas; and higher impedance, balanced inputs of quad antennas.

The operation of these single-core, higher order (e.g., trifilar, quadrifilar) winding transformers can be briefly explained using the 1:2.25 unun shown in Figure 7-1. The high frequency schematic of Figure 7-1a assumes that the characteristic impedance of the two transmission lines (the center winding being common) is at the optimized value of $R_L/3$ and that the reactance of the coiled transmission line is much greater than $R_L/3$. Figure 7-1a shows the output voltage to be the sum of two direct voltages of $V_{in}/2$ and one delayed voltage of $V_{in}/2$. This leads, in the mid-band, to $V_{out} = 3/2\ V_{in}$ and therefore an impedance ratio of 1:2.25.

The top winding carries 2/3 *I* and the bottom two windings 1/3 *I*. The output winding is shown as the top winding in the figure, and this is the configuration with which Sevick experimented. This configuration will operate as a 1:2.25 transformer, but it disrupts operation of the windings for TEM wave propagation. This configuration is more closely a simple magnetically coupled transformer. Sevick also found, as mentioned later, that improved performance is obtained by taking the output from the center wire to maintain balance.

Since the windings of the transformer are in series aiding (100% mutual coupling; Figure 7-1b), fewer trifilar turns are required compared with the bifilar (1:4) model, which gives the same low frequency response as the transformer in Figure 4-5a. Thus, with fewer turns and shorter transmission lines, these transformers are capable of much higher frequency responses than the regular 1:4 Ruthroff transformer.

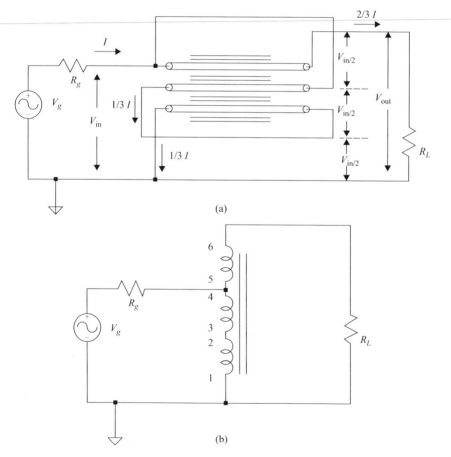

Figure 7-1   *Schematic of mid- and low-band operation of a trifilar transformer for 1:1.5 operation.*

The high frequency improvement is on the order of 2.25 times. In a way, this transformer is a combination of Ruthroff and Guanella's methods. Two-thirds of the output voltage is tied directly to the input (raised by its own bootstrap), and the other one-third is the delayed output from the output transmission line.

The output winding can also be tapped easily, giving a broadband impedance ratio near 1:2. The analysis of a tapped winding is more difficult. Operation at the low frequency end is obvious because it behaves as a magnetic autotransformer. Operation as a transmission line transformer is more difficult because the fields associated with the other windings interact with the field of the tapped winding and its neighbors. Analysis without electromagnetic modeling software is impossible. Nonetheless, Sevick produced experimental results showing that these transformers operate efficiently over a useable frequency range. The same can be said of the multiple conductor transformers. The transmission line aspect of the windings is severely disrupted in the four- and five-conductor winding transformers. It is more correct to analyze these transformers as magnetic transformers with very closely coupled windings.

The process of using multiple wires can be continued further. For example, a quadrifilar winding has an output voltage of $V_{out} = 4/3\ V_{in}$ with an impedance ratio of 1:1.78. A quintufilar winding creates an output voltage of $V_{out} = 5/4\ V_{in}$, yielding an impedance ratio of 1:1.56. In fact, Sevick successfully constructed a seven-winding (septufilar) transformer that gave a very broadband ratio of 1:1.36. It can also be shown that for the same low frequency response as the regular 1:4 Ruthroff transformer, the high frequency responses of the quadrifilar and quintufilar trans-formers are better by factors of about four and five, respectively. In general, the greater the number of windings, the shorter the transmission lines, the more the in-phase voltages are summed, and the greater the high frequency response.

This chapter presents many low-ratio designs that use bifilar and higher order windings. Various kinds of conductors, such as rectangular line and coax cable, are also investigated. The presentation on transformers with more than two windings is unique and should find many uses. Also, a novel concept is presented for rearran-ging windings for optimally matching 50 Ω to higher or lower impedances.

7.2   1:1.5 Ununs

There are many ways of obtaining an impedance ratio near 1:1.5 in an unun transformer using the Ruthroff method of summing a direct voltage with a delayed voltage (or voltages). In the bifilar case, the top winding is tapped at the appropriate point, yielding an output voltage (V_{out}) nearly equal to 1.25 V_{in}, where V_{in} is the input voltage. Higher order windings, such as trifilar and quadrifilar, can also have their top winding appropriately tapped for the desired 1.25 V_{in} output voltage. Even though trifilar and quadrifilar transformers are better than bifilar transformers, the best choice, by far, is the quintufilar transformer. This transformer has an impe-dance ratio of 1:1.56 without tapping and possesses a wider bandwidth than the other three. This section treats the tapped bifilar and quintufilar cases.

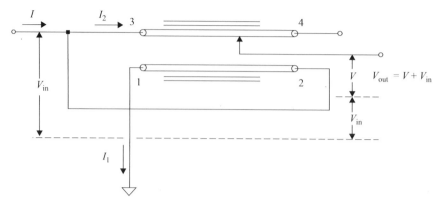

Figure 7-2   Schematic for analysis of tapped bifilar Ruthroff transformer.

7.2.1   Tapped Bifilar Transformers

The model for the analysis is shown in Figure 7-2. With terminal 3 connected to terminal 2, a gradient of V_{in} exists across the length of the bottom winding. The same potential gradient will exist across the top winding. If the transmission line is wound on a rod, then the tapped voltage (V), which is calculated from terminal 3, is

$$V = V_{in}l/L \tag{7-1}$$

where

    L = the total length of the wire from terminal 3 to terminal 4
    l = the length from terminal 3 to the tap

Thus, the output voltage (V_{out}) becomes

$$V_{out} = V_{in}(1 + l/L) \tag{7-2}$$

and the impedance transformation ratio (ρ) is

$$\rho = (V_{out}/V_{in})^2 = (1 + l/L)^2 \tag{7-3}$$

When $l = L$, ρ has the familiar value of 1:4. Since $V_{in} \times I = V_{out} \times I_2$, the output current becomes

$$I_2 = I/(1 + l/L) \tag{7-4}$$

With a toroidal core, each turn encloses all of the flux, so the gradient can be tapped only at integral turns. Thus, for a toroid, equation (7-3) becomes

$$\rho = (1 + n/N)^2 \tag{7-5}$$

where

    N = the total number of turns
    n = the integral number of turns counted from terminal 3

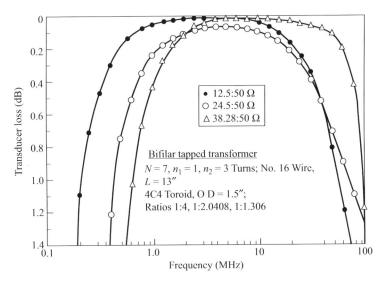

Figure 7-3   *The plots show the frequency response of a bifilar tapped transformer at 3 different impedance levels.*

Figure 7-3 shows the loss as a function of frequency for a tapped bifilar transformer when the output is terminated in 50 Ω. The transformer was built with seven turns of no. 16 wire on a 1 1/2 in OD core of 4C4 material ($\mu = 125$). The top winding is tapped at one and three turns from terminal 3. With an output also at terminal 4, the impedance ratios available are 1:1.31, 1:2.04, and 1:4, respectively. Measurements on other tapped bifilar transformers exhibited similar results.

Figure 7-4 shows two practical tapped bifilar unun transformers with impedance ratios near 1:1.5. The coax connectors are at the low impedance sides with the following parameters:

Left: Eleven bifilar turns of no. 14 wire on a 1 1/2 in OD, 52 toroid ($\mu = 250$). The tap is two turns from terminal 3 in Figure 7-2. The impedance ratio is 1:1.4. At the 38:50 Ω impedance level (which is near optimum), the recommended range is from 1.5 to 15 MHz. Above 15 MHz, the ratio increases and becomes a complex quantity. The power rating is 1 kW continuous power.

Right: Fourteen bifilar turns of no. 16 wire on a 1/2 in diameter rod of no. 61 material ($\mu = 125$). The tap is 2 1/4 turns from terminal 3 in Figure 7-2. The impedance ratio is 1:1.35. At the 37:50 Ω impedance level (which is near optimum), the recommended range of operation is from 3.5 to 10 MHz. Above 10 MHz, the ratio decreases and becomes a complex quantity. The power rating is 500 W continuous power. This transformer is considerably poorer than the toroidal version in Figure 7-4.

Figure 7-5 shows an efficiency curve that tapped bifilar transformers generally exhibit. The lowest efficiency appears at about the 1:2.25 ratio. From this curve, it is apparent that acceptable operation occurs at impedance ratios less than 1:1.5 and

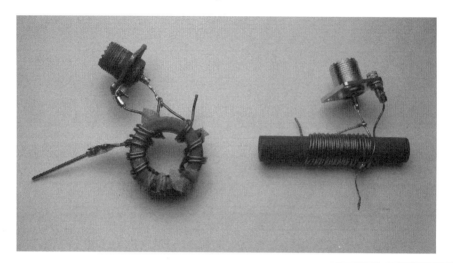

Figure 7-4   Photo showing two practical tapped bifilar unun transformers with impedance ratios near 1:1.5.

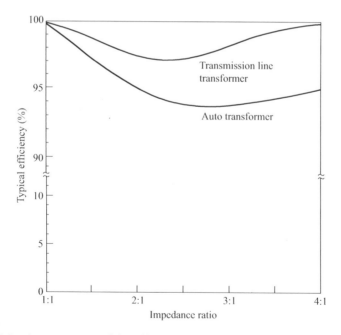

Figure 7-5   A comparison of the efficiency between tapped bifilar transmission line transformers and an autotransformer. Both transformers use no. 16 wire on 1 1/2 in 4C4 cores.

greater than 1:3. Also evident is the much poorer performance of the auto-transformer compared with the transmission line transformer.

Several conclusions can be made concerning tapped bifilar transformers:

1. For the 1:2.04 connection, the loss is considerably greater and the bandwidth considerably less than the other two. The greater loss (also shown in Figure 7-5) suggests that, to a large degree, conventional transformer action is taking place. The transformer should not be operated in this mode. A much better way for obtaining impedance ratios of 1:2 is described in section 7.3.

2. The high frequency response for the 1:1.31 connection is considerably greater than for that of the 1:4 connection. The effective length of the transmission line is shorter, and the characteristic impedance of a transmission line using no. 16 wire favors the 38.28:50 Ω level. For many applications, low impedance ratios (around 1:1.3), obtained by tapping, are practical; high efficiencies can be obtained. If even greater bandwidths at these low impedance ratios are desired, then going to higher orders of windings is recommended (section 7.2.2). These higher order winding transformers can also be tapped for much better band-width than the bifilar transformer.

3. Low impedance ratios in the area of 1:1.3–1:1.5, with tapped bifilar transfor-mers, require considerably more bifilar turns (for the same low frequency response) than ratios of 1:3–1:4. This is because only a small part of the turns in the top winding in Figure 7-2 play a role in the low frequency model.

7.2.2 Quintufilar Transformers

Quintufilar transformers, although being somewhat more difficult to construct, are by far the superior transformer for use around the 1:1.5 impedance ratio. Figure 7-6 shows pictorials for the rod and toroidal versions. A definite pattern is visible, which should help in connecting the various numbers given in the schematics that follow. The photographs of the practical designs will also help.

Figure 7-7 shows schematic diagrams for two versions of the quintufilar, unun transformer with an impedance ratio of 1:1.56. Figure 7-7a, using no. 14 or 16 wire wound tightly on a rod (as in Figure 7-6b), gives its highest frequency response

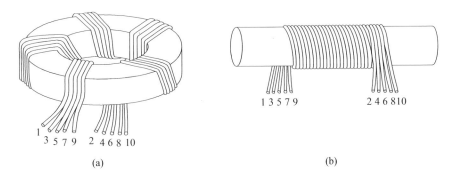

Figure 7-6 Pictorial shows quintufilar transformer construction.

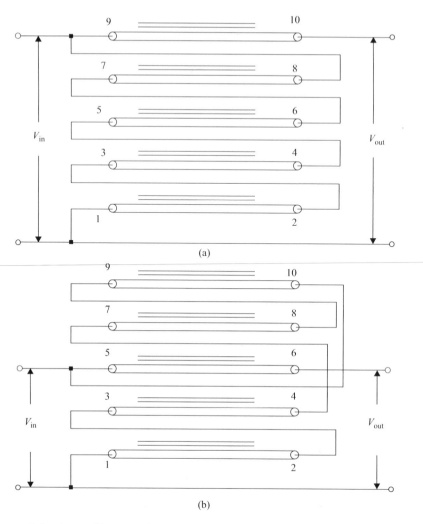

Figure 7-7   *Quintufilar transformers with impedance ratios of 1:1.56: (a) High impedance operation. (b) Windings configured for low impedance operation.*

when matching 32 to 50 Ω. On a toroid, which cannot be wound as tightly as on a rod, the windings favor matching 50 to 78 Ω. Instead of taking the output from the top wire as in Figure 7-7a, but using the center wire as in Figure 7-7b, the characteristic impedance is considerably lowered, and optimum performance with a toroidal core can easily occur at the 32:50 Ω level. All of the optimum impedance levels can be easily increased by appropriate thickness of insulation on the wires. And finally, it should be noted that these transformers with low impedance transformation ratios, particularly using higher order windings such as quadrifilar and quintufilar, become quite bilateral in nature. They can be used both as a step-up or a step-down transformer. The main difference is that their high frequency response is

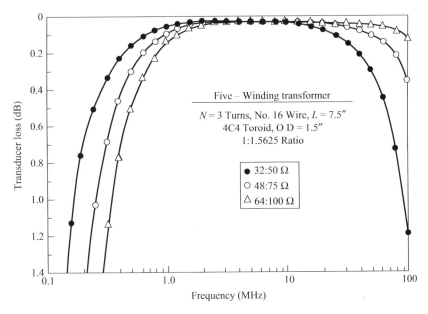

Figure 7-8   Loss versus frequency for a five-winding transformer at various impedance levels.

generally twice as good in the favored direction for which they were designed. In practice, this means a constant transformation ratio up to 45 to 60 MHz in one direction and 25 to 30 MHz in the other.

The loss as a function of frequency for a five-winding transformer is illustrated in Figure 7-8. The best high frequency response occurs at the 64:100 Ω level. Here, a loss of only 0.1 dB extends from 1.2 to 90 MHz! The performance at the 48:75 Ω level also shows very good high frequency response. For best performance at the 32:50 Ω level, the schematic of Figure 7-7b should be employed.

Figure 7-9 is a photograph of four 1:1.56 rod ununs using the schematic in Figure 7-7a. The one on the bottom left is constructed by first forming a five-conductor ribbon, held in place every 3/4 in by a 3M no. 92 clamp. The others are constructed by simply adding one winding at a time. The transformer on the bottom left is specifically designed to match 50 to 78 Ω. At this impedance level, the transformation ratio is constant from 1.5 to 40 MHz. At the impedance level of 32:50 Ω, it is constant from 1.5 to 20 MHz. The other three transformers are designed for the 32:50 Ω level. At this level, their impedance ratios are constant from 1.5 to 40 MHz. At the 50:78 Ω level, they are constant from 1.5 to 20 MHz.

The parameters for these four transformers, which have the cable connectors at the low impedance side, are as follows:

Bottom left: Nine quintufilar turns on a 1/2 in diameter rod ($\mu = 125$). The top winding (Figure 7-7A) is no. 14 wire covered with 10 mil wall plastic tubing. The other four windings are no. 16 wire. The power rating is 1 kW continuous power.

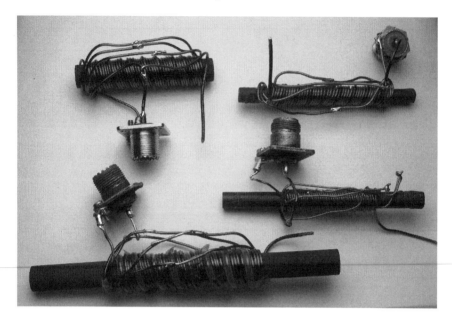

Figure 7-9   Photo of four quintufilar rod ununs. The transformer on the bottom left is designed to match 50 to 78 Ω. The other three are designed to match 50 to 32 Ω.

Top left: Nine quintufilar turns on a 1/2 in diameter rod ($\mu = 125$). The top winding (Figure 7-7A) is no. 14 wire, and the other four are no. 16 wire. The power rating is 1 kW continuous power.

Top right: Nine quintufilar turns on a 3/8 in rod ($\mu = 125$). The top winding (Figure 7-7A) is no. 14 wire, and the other four are no. 16 wire. The power rating is also 1 kW continuous power.

Bottom right: Nine quintulilar turns on a 3/8 in diameter rod ($\mu = 125$). The top winding (Figure 7-7A) is no. 16 wire, and the other four are no. 18 wire. The power rating is 200 W continuous power. This transformer can handle 800 W continuous power with only a small temperature rise.

Toroidal transformers have an advantage over their rod counterparts because they have a closed magnetic path and benefit from higher permeability ferrites. Rod transformers are insensitive to permeability. As a result, fewer turns can be used with toroidal transformers for the same low frequency response. This translates to shorter transmission lines and higher frequency responses. Figure 7-10 is a photograph of two 1:1.56 toroidal ununs. The transformers have small cores with high permeabilities. Their bandwidths are about equal to the four rod transformers in Figure 7-9. From an overall power and bandwidth capability, transformers similar to the one on the left in Figure 7-10 are preferred for low impedance ratios at the 1 kW power level. Toroids with outside diameters of 1 1/2 to 1 3/4 in and permeability of 250–300 allow enough space to wind the appropriate sizes of conductors. Further, this permeability range still offers efficiencies in the 98–99% region.

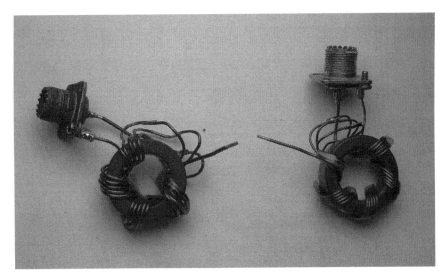

Figure 7-10   *The left transformer is a quintufilar design using the schematic of Figure 7-7b. The transformer on the right implements a modified version of the schematic of Figure 7-7a.*

The parameters for the transformers in Figure 7-10 are as follows:

Left: Four quintufilar turns on a 1 1/2 in OD, no. 64 toroid ($\mu = 250$). The schematic in Figure 7-7B is used. The center coil, winding 5–6, has no. 14 wire, and the other four have no. 16 wire. When matching at the 32:50 Ω level, the impedance ratio is constant from 1.5 to 50 MHz. At the 50:78 Ω level, it is constant from 1.5 to 25 MHz. Since power ratings are more dependent on conductor sizes and not core sizes, this smaller transformer has the same 1 kW continuous power rating as ones built on larger 2.4 in cores. Further, it is an excellent bilateral transformer—stepping down or up from 50 Ω.

Right: Five quintufilar turns on a 1 1/2 in OD, 52 toroid ($\mu = 250$). The schematic consists of interleaving winding 9–10, in Figure 7-7a, between winding 7–8 and winding 5–6. This tends to be a good compromise for a quasi-bilateral transformer. Winding 9–10 has no. 16 wire, and the other four have no. 18. When matching 50 to 78 Ω, the constant impedance ratio is from 1.5 to 100 MHz. At the 32:50 Ω level, it is constant from 1.5 to 50 MHz. A conservative power rating is 200 W continuous power. This transformer has operated at 700 W without failure.

And, finally, Figure 7-11 shows a photograph of three other 1:1.56 unun transformers that might prove useful to some readers. The transformer on the left is designed to match (nominally) 100 to 150 Ω. The transformer in the center is designed to match 32 to 50 Ω from 1.5 to 150 MHz. The transformer on the right, a coax cable version whose schematic is shown in Figure 7-12, matches 32 to 50 Ω with 5 kW capability. All connectors are on the low impedance side of the transformers.

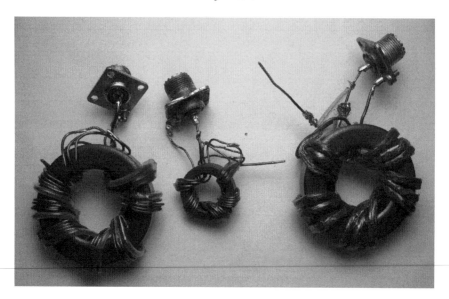

*Figure 7-11   The transformer on the left matches 100 to 150 Ω. The center
transformer matches 50 to 32 Ω and has response 150 MHz wide.
The transformer on the right uses low impedance cable and has 5 kW
capability.*

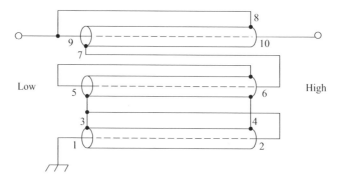

*Figure 7-12   This is the coax cable version of the transformer of Figure 7-7a. The
outer braid of the bottom two cables are connected together.*

Their specific parameters are as follows:

Left: Eight quintufilar turns on a 2.4 in OD, no. 67 toroid ($\mu = 40$). The schematic in Figure 7-7a is used. Winding 9–10 uses no. 14 wire covered with a
20 mil wall Teflon tubing. The other four windings are no. 16 wire. At the
100:156 Ω level, the impedance ratio is constant from 1.5 to 30 MHz. The
power rating is at least 1 kW continuous power. For a much higher frequency
response, five or six quintufilar turns on a higher permeability (250 to 290)
and smaller outside diameter (1 3/4 to 2 in) toroid are recommended.

Center: Five quintufilar turns on a 3/4 in OD, no. 64 toroid ($\mu = 250$). The schematic in Figure 7-7b is used. Winding 5–6 uses no. 16 wire, and the other four use no. 18 wire. At the 32:50 Ω level, the impedance ratio is constant from 1.5 to 150 MHz. At the 50:78 Ω level, it is constant from 1.5 to 75 MHz. The reasons for this very wideband capability are the use of a relatively high permeability ferrite and the short length of transmission lines—only 5 1/2 in long. A conservative power rating is 200 W continuous power. This very small transformer has also operated at 700 W.

Right: Five quintufilar turns of three coax cables connected as a five-winding transformer (Figure 7-12) on a 2.4 in OD, 52 toroid ($\mu = 250$). The top coax cable has a no. 14 inner conductor with two layers of 3M no. 92 tape. The outer braid is from RG-58/U cable and is tightly wrapped with 3M no. 92 tape. The characteristic impedance of this coax is 14 Ω. The other two coax cables use no. 16 wire for inner conductors with two layers of 3M no. 92 tape. The outer braids, from RG-58/U cable, are also wrapped with 3M no. 92 tape. Their characteristic impedances are 19.5 Ω. The highest frequency response occurs at the 45:70 Ω level, where the transformation ratio is constant from 1 to 40 MHz. At the 32:50 Ω level, the ratio is constant from 1 MHz to 30 MHz. Since the current is distributed evenly about the inner conductor of the coax cables, a conservative power rating is 5 kW continuous power. If ML or H Imideze wire were used, the voltage breakdown of this transformer would rival that of RG-8/U cable. Further, by substituting no. 12 wire for the no. 14 wire and no. 14 wire for the no. 16 wire, a more favorable impedance level would be 32:50 Ω. Also, the power rating would improve at least twofold.

7.3   1:2 Ununs

This section presents two methods for obtaining transformation ratios around 1:2 related to Ruthroff's technique of adding direct voltages to voltages that have traversed coiled transmission lines: (1) a tapped trifilar winding, yielding ratios of 1:2.25 and 1:2; and (2) a quadrifilar winding, yielding a ratio of 1:1.78. Both result in very broadband performances since they can employ rather short transmission lines and still satisfy the low frequency requirements.

Figure 7-1a displays the circuit diagram for a trifilar design that produces very broadband operation at a 1:2.25 transformation ratio. By connecting the top of the load resistor (R_L) to a tap on the top winding 5–6, an equally broadband (and efficient) ratio can be obtained near 1:2. The tapped version is shown in Figure 7-13. This trifilar transformer is greatly superior to the tapped bifilar transformer in obtaining ratios near 1:2.

Figure 7-14 clearly shows the very high efficiency of the trifilar 1:2.25 (untapped) transformer. Losses less than 0.1 dB extend over wide bands for three different impedance levels. Transformers with no. 14 wire on toroidal cores have their maximum high frequency response near the 44.44:100 Ω level. In this case, the loss is less than 0.04 dB from 1.2 to 30 MHz. This transformer (explained in Chapter 8) also has a 1:9 connection.

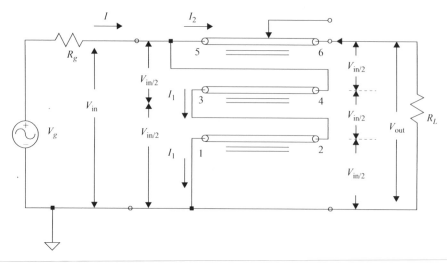

Figure 7-13   The schematic of a tapped trifilar transformer shows a connection for 1:2.25. If the load is connected to the tap instead, it provides a ratio between 1:1 and 1:2.25 depending on the relative position of the tap.

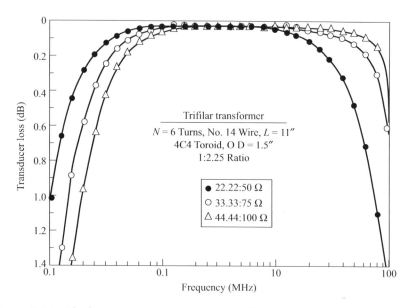

Figure 7-14   The loss versus frequency for a trifilar transformer configured for 1:2.25 at three different impedance levels.

No. 14 wire yielded the best high frequency performance at the 44.44:100 Ω level for the untapped trifilar connection shown in Figure 7-14. Therefore, an investigation was undertaken with transmission lines of lower characteristic impedances to obtain better performance at lower impedance levels. A six-turn

Figure 7-15   A photo of the 7/64 in rectangular line trifilar transformer.

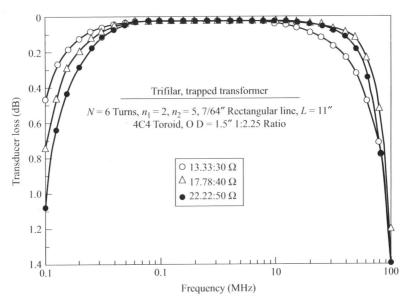

Figure 7-16   These plots show the response of a 7/64 in rectangular line versus impedance ratio.

trifilar transformer using 7/64 in rectangular line, with insulation of 3M no. 92 (2.8 mil thick), was wound on a 4C4, 1 1/2 in OD toroid (Figure 7-15). Figure 7-16 shows the performance for the 1:2.25 ratio. As shown, the response at the 22.22:50 Ω level is much better than that of the no. 14 windings shown in Figure 7-14.

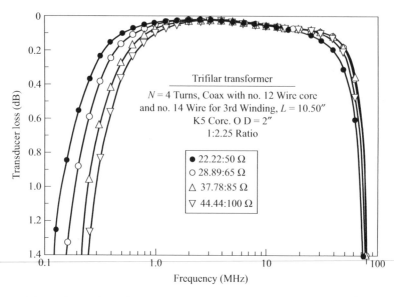

Figure 7-17   These plots show frequency response versus impedance level for a transformer built with very low impedance coax.

The 0.04 dB loss extends from 0.5 to 30 MHz. It is also evident that this 7/64 in trifilar rectangular line is optimum at the 17.78:40 Ω impedance level. This transformer will also be explained further in Chapter 8.

Investigations of two other transmission line configurations revealed interesting results. One used a low impedance coax cable for the top two windings in Figure 7-14. The inner conductor of no. 12 wire was insulated with two layers of 3M no. 92 tape. The bottom winding used an insulated no. 14 wire. The performance curves are shown in Figure 7-17. Notice that the performance at the 22.22:50 Ω level compares quite favorably to the 44.44:100 Ω level of the no. 14 wire transformer. The slight increase in loss above 5 MHz is characteristic of K5 material.

The second transformer used 7/64 in rectangular line with 3M no. 92 insulation for the top two windings. The bottom winding was an insulated no.16 wire. The performance at the 22.22:50 Ω level (Figure 7-18) also compares favorably to the 44.44:100 Ω level of the no. 14 wire transformer. Thus, it can be concluded that the characteristic impedances of the top two windings in Figure 7-13 are most important in determining the high frequency performance of a trifilar bootstrap transformer.

Probably the most interesting technique for obtaining improved low impedance operation of a trifilar wire transformer with a toroidal core is the rearrangement of the windings shown in Figure 7-19. The top winding in Figure 7-13 is simply placed in the middle, where it carries the larger current and is closely coupled to both of the other two windings. This lowers the characteristic impedance and

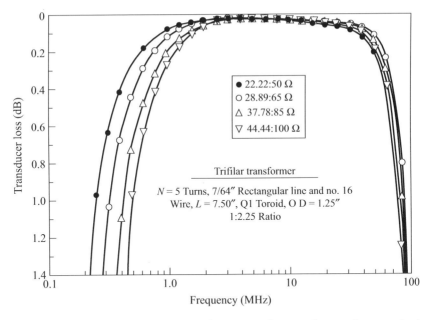

Figure 7-18 *These plots show the performance of a transformer that uses both a rectangular line and a no. 16 wire to produce a trifilar transformer.*

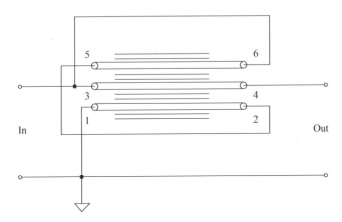

Figure 7-19 *The schematic of a 1:2.25 toroidal transformer that shows optimum performance at the 22.22:50 Ω level using no. 14 wire.*

improves the low impedance performance. The configuration creates a symmetric TEM flow on the two parallel transmission lines. Actual data show that a transformer similar to the one in Figure 7-13, but with the top conductor placed in the center as in Figure 7-19, has a high frequency response at the 22.22:50 Ω level similar to the response at the 44.44:100 Ω level in Figure 7-13. Thus, when stepping up from 44.44 Ω (also 50 Ω) to 100 Ω, Figure 7-13 is preferred. When stepping down from 50 to 22.22 Ω (25 Ω), Figure 7-19 is preferred. Windings of no. 16 wire gave similar results. However, the tightly wound rod transformer presents a completely different case: rod transformers favor Figure 7-13 when matching 22.22 Ω to 50 Ω.

An analysis of the tapped trifilar transformer for determining the impedance transformation ratio (ρ) is similar to the bifilar case. For the trifilar rod transformer, the output voltage becomes

$$V_{out} = V_{in} + V_{in}/2 \times l/L \tag{7-6}$$

where

   l = the length of the transmission line from terminal 5 in Figure 7-14 or from terminal 3 in Figure 7-19
   L = the total length of the transmission line

The impedance ratio then becomes

$$\rho = (V_{out}/V_{in})^2 = (1 + l/2L)^2 \tag{7-7}$$

When $l = L$, the ratio of 1:2.25 is obtained.

For the toroidal transformer, where only integral turns are effective, equation (7-7) is

$$\rho = (1 + n/2N)^2 \tag{7-8}$$

where

   n = the number of turns from terminal 5 in Figure 7-13 or from terminal 3 in Figure 7-19
   N = the total number of trifilar turns

Figure 7-20 shows the outstanding performance of a toroidal transformer tapped at about the 1:2 impedance ratio. The tap is at five of six turns from terminal 5 diagrammed in Figure 7-13. As shown, using no. 14 wire optimizes at the 50:100 Ω level, where the loss is less than 0.1 dB from 750 kHz to 75 MHz and less than 0.04 dB from 1 MHz to 40 MHz. A similar transformer using the diagram in Figure 7-19 gives about the same results but at the 25:50 Ω level. These two transformers are excellent 1:2 transformers for matching 50 Ω to 100 Ω (Figure 7-13) or 50 Ω to 25 Ω (Figure 7-19).

The trifilar transformer using rectangular line also had taps at two and five turns from terminal 5 (Figure 7-13). As mentioned, this transformer was designed for low impedance operation and had the highest frequency response at the 17.78:40 Ω impedance level when operating as a 1:2.25 transformer. With the three output ports, impedance ratios of 1:1.36, 1:2.01, and 1:2.25 are available. Figure 7-21 shows the performance curves for the three different ratios when the high side is terminated in 50 Ω. True transmission line transformer performance is obtained with the taps at two and five turns. Taps at one, three, and four turns gave

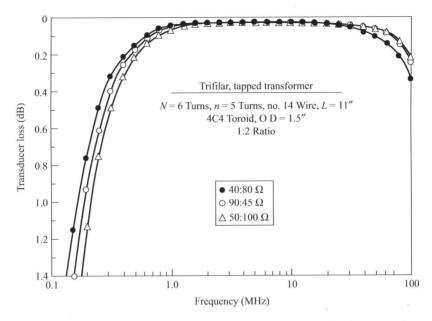

Figure 7-20   These plots show loss vs. frequency for a tapped trifilar transformer at 1:2 impedance ratio and at three different impedance levels.

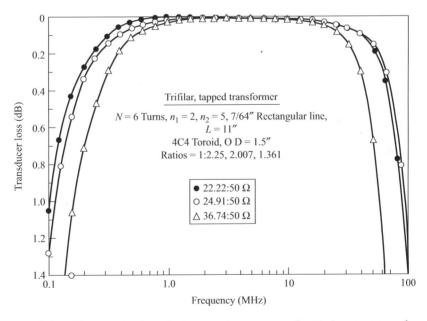

Figure 7-21   These plots show the loss vs. frequency at the 50 Ω output impedance level for a tapped trifilar rectangular line transformer.

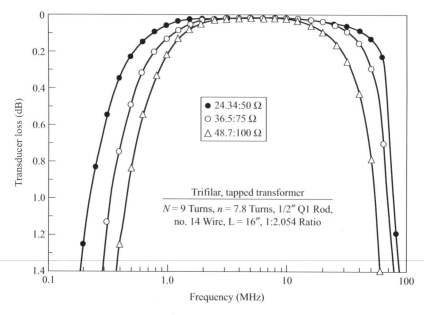

*Figure 7-22   These plots show the loss vs. frequency for a trapped trifilar
transformer using no. 14 wire on a ½ inch diameter Q1 rod at
three different impedance levels.*

similar results. The figure illustrates that the high frequency performance of the
1:1.36 ratio was not as good as for the other two. This is because the characteristic
impedance of the rectangular line is less than optimum for the 36.74:50 Ω level.
In fact, a similar tapped trifilar transformer using no. 14 wire is optimum at the
36.74:50 Ω level.

Figure 7-22 presents the results of a trifilar transformer with nine tightly
wound turns of no. 14 wire on a 1/2 in diameter Q1 rod ($\mu = 125$) and tapped at 7.8
turns from terminal 5 (Figure 7-13), giving a transformation ratio of 1:2.05. The rod
was 3 3/4 in long. Optimum high frequency performance occurs at the 24.34:50 Ω
level. Because of the tight winding, rod transformers using the schematics in
Figure 7-13 or Figure 7-19 gave similar results. Although the rod transformer is not
quite as good as a toroidal one of similar permeability (Figure 7-21), excellent
bandwidth at high efficiency is obtained. At the 24.34:50 Ω level (Figure 7-22), the
loss is less than 0.1 dB from 800 kHz to 45 MHz and less than 0.04 dB from 2 to
25 MHz. This transformer also has transformation ratios of 1:4 and higher and will
be described in greater detail in Chapter 8.

Figure 7-23 shows a photograph of three tapped trifilar rod transformers with
impedance ratios of 1:2 and 1:2.25. They are all designed to handle 1 kW continuous power. The top transformer is designed to match 50 Ω to 100 Ω or 112.5 Ω.
The bottom two transformers are designed to match 50 Ω to 22.22 Ω or 25 Ω.

Figure 7-23   *This photo shows three tapped trifilar rod transformers with impedances of 1:2 and 1:2.25. The top transformer is designed to match 50 Ω to 100 Ω or 112.5 Ω. The bottom two transformers are designed to match 50 Ω to 22.22 Ω or 25 Ω.*

The cable connectors are all on the low impedance sides of the transformers. Their parameters are as follows:

Top: Ten trifilar turns on a 1/2 in diameter, no. 61 rod ($\mu = 125$). The top winding is no. 14 wire, insulated with a 20 mil wall Teflon tubing, and is tapped at 7 3/4 turns from terminal 5 in Figure 7-13. The other two windings are no. 16 wire. The two impedance ratios of 1:2 (the tapped output) and 1:2.25 arc constant from 3.5 to 30 MHz. If operation in the 160 m band is also desired, then 12 trifilar turns with a tap at 10 turns is recommended.

Middle: Ten trifilar turns on a 1/2 in diameter, no. 61 rod ($\mu = 125$). The top winding is no. 14 wire and is tapped at eight turns from terminal 5 in Figure 7-13. The other two windings are no. 16 wire. The two impedance ratios of 1:2 (the tapped output) and 1:2.25 are constant from 1.5 to 45 MHz. The three wires were wound as a ribbon held in place by clamps of 3M no. 92 tape every 3/4 in. This also gave a little spacing between trifilar turns. This

Figure 7-24   This photo shows two trifilar transformers that are tapped to give 1:2 transformation ratio.

optimized the performance at the 22.22:50 Ω and 25:50 Ω levels. This same technique was applied to the other two rod transformers in Figure 7-24.

Bottom: Thirteen trifilar turns on a 3/8 in diameter, no. 61 rod ($\mu = 125$). The top winding is no. 14 wire and is tapped 11 turns from terminal 5 in Figure 7-13. The other two windings are no. 16 wire. The impedance ratios of 1:2 (the tapped output) and 1:2.25 are constant from 1.5 to 4.5 MHz. As mentioned already, the three wires are wound as a ribbon held together by sections of 3M no. 92 tape. Performance slightly favors this transformer over its 1/2 in counterpart.

Figure 7-24 is a photograph of two tapped trifilar toroidal transformers. The one on the left is made up of two low impedance coax cables with their outer braids connected in parallel and acting as the third conductor. The schematic is shown in Figure 7-25. This transformer, which is optimized to match 50 Ω to 22.22 Ω or 25 Ω, is conservatively rated at 5 kW continuous power. Also, since it uses four layers of 3M no. 92 tape as the insulation on the no. 14 inner conductor, its voltage breakdown is similar to that of RG-8/U cable. Further, it demonstrates that the inner conductor of a coax cable in a bootstrap-connected transformer possesses a potential gradient which can be tapped. The transformer on the right is included here since many readers have the popular 2.4 in OD toroid ($\mu = 125$) and would find it convenient to use. The parameters for these two transformers are as follows:

Left: Six trifilar turns of low impedance coax ($Z_0 = 18.5$ Ω) on a 1 1/2 in OD, K5 toroid ($\mu = 290$). The no. 14 inner conductors have four layers of 3M no. 92 tape. The outer braids, which came from RG-58/U cable, are also wrapped with 3M no. 92 tape. The inner conductor of the top coax in

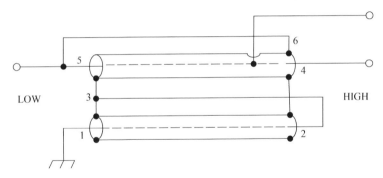

Figure 7-25   *The schematic shows a tapped bifilar coaxial transformer to produce a ratio between 1:1 and 1:2.25. The braid acts as the third wire of a trifilar transformer at low frequencies.*

Figure 7-24 is tapped at five turns from terminal 5. When matching 50 Ω to 22.22 Ω or to 25 Ω, the impedance ratio is constant from 1 to over 50 MHz. *Right*: Nine trifilar turns of no. 14 wire on a 2.4 in OD, no. 61 toroid ($\mu = 125$). The top winding in Figure 7-13 is tapped at eight turns from terminal 5. This transformer matches 100 Ω to 44.44 Ω or 50 Ω with a constant impedance ratio from 1.5 to 40 MHz. The power rating is a conservative 1 kW continuous power.

Figure 7-26 shows the schematics for two quadrifilar transformers, which yield very broad transformation ratios of 1:1.78. The transformer in Figure 7-26b, because of winding 5–6 being interleaved between windings 7–8 and 3–4, operates better at low impedance levels with toroidal cores. As with trifilar transformers, quadrifilar transformers can also be tapped to give excellent performance at ratios less than 1:1.78. This can best be seen from Figure 7-26a. For a tapped quadrifilar transformer, the output voltage is

$$V_{\text{out}} = V_{\text{in}}(1 + n/3N) \tag{7-9}$$

for a toroidal transformer where

$n =$ the number of turns from terminal 7 in Figure 7-26a and from terminal 5 in Figure 7-26b

$N =$ the number of trifilar turns

The transformation ratio (ρ) is then

$$\rho = (V_{\text{out}}/V_{\text{in}})^2 = (1 + n/3N)^2 \tag{7-10}$$

Figure 7-27 is a photograph of two quadrifilar transformers with impedance ratios of 1:1.78. They both employ Figure 7-26b, which favors matching 28 Ω to 50 Ω. Their parameters are as follows:

Left: Six quadrifilar turns on a 1 1/2 in OD, 250L toroid ($\mu = 250$). Winding 5–6 (Figure 7-26b) uses no. 12 H Imideze wire covered with two layers of 3M no. 92 tape. The other three windings are no. 14 H Imideze wire. At the optimum

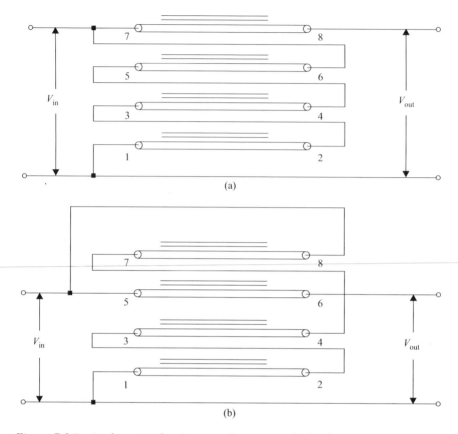

*Figure 7-26   A schematic showing two alternate methods of creating a 1:1.78
transformation ratio. (a) shows windings for a high impedance
application, and (b) shows windings for a low impedance application.*

impedance level of 37:65 Ω, the impedance ratio is constant from 1 to
40 MHz. When matching 50 Ω to 89 Ω (step-up) or 50 Ω to 28 Ω (step-
down), the impedance ratio is constant from 1 to 30 MHz. This rather husky
transformer, which should be able to handle 2 to 3 kW continuous power, is
quite bilateral.

Right: Five quadrifilar turns of no. 14 wire on a 1 1/2 in OD, 4C4 toroid ($\mu =$
125). This transformer, which uses Figure 7-26b, is optimized to match 50 Ω
to 28 Ω. At this impedance level, the transformation ratio is constant from 1 to
over 40 MHz. When matching 50 Ω to 89 Ω, the impedance ratio is constant
from 1.5 to 20 MHz. A conservative power rating is 1 kW continuous power.

7.4   1:3 Ununs

In many cases the 1:2.25 unun or the 1:4 unun can provide adequate matching when
a 1:3 impedance ratio exists. This is particularly true when matching into antennas

Figure 7-27   *This photo shows examples of two quadrifilar 1:1.78 ratio*
transformers. They both implement the schematic of Figure 7-26(b)
which favors matching 28 Ω to 50 Ω. The left transformer is rated at
2 kW because of the use of no. 12 wire. The one on the right is rated
at 1 kW continuous power.

where the radiation resistance varies with frequency. An example is the resonant quarter-wavelength vertical antenna over a lossless ground system. It has a resistive input impedance of 35 Ω for any reasonable thickness of antenna. This results in a VSWR, at resonance, of 1.4 when feeding directly with 50 Ω coax cable. But the lowest VSWR of about 1.25 occurs a little higher in frequency because of the increased radiation resistance. Resonance and lowest VSWR occur at the same frequency only when the antenna's resonant impedance is 50 Ω (if 50 Ω coax is used) or when a transformer or network matches the impedance of the transmission line to the impedance of the antenna. But on occasion a 1:3 unun is desirable. An example is the 1:12 balun, shown in Chapter 9, which matches 50 Ω unbalanced to 600 Ω balanced. This balun uses a 1:3 unun in series with a 1:4 balun. Many solid-state circuits, which are critical of impedance levels, could also find a 1:3 unun of value.

Sevick investigated two methods for obtaining ununs with impedance transformation ratios at or about 1:3. One uses the tapped bifilar schematic shown in Figure 7-2, and the other employs a quintufilar schematic similar to those in Figure 7-8. In the quintufilar case, the windings are interleaved and the input connections (on the left side) are made to result in two broadband ratios of 1:1.56 and 1:2.78. These transformers are described in Chapter 10, which deals with multimatch transformers. This section describes the tapped bifilar case. Figure 7-28 shows a photograph of a tapped bifilar rod transformer and a tapped bifilar toroidal transformer. Each has three taps and therefore four different transformation ratios. They are both capable of handling 1 kW continuous power. The cable connectors

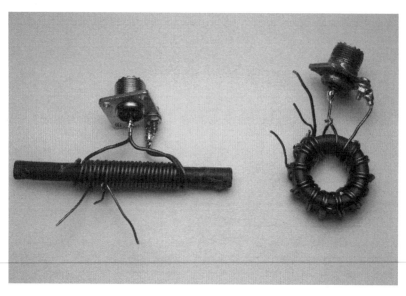

Figure 7-28   This photo shows examples of two tapped bifilar transformers with impedance ratios approximating 1:1.3. Each transformer has taps between 1:3.2 and 1:4.

are on the low impedance sides. The parameters for these two transformers are as follows:

Left: Sixteen bifilar turns of no. 14 wire on a 3/8 in diameter, no. 61 rod ($\mu =$ 125). The top winding (from Figure 7-2) is tapped at 9 3/4, 10 7/8, and 12 turns from terminal 3. Sevick tried for 10, 11, and 12 turns but wound up a little short on the first two. These differences in turns are insignificant. The impedance ratios are 1:3.2, 1:3.5, 1:3.75, and 1:4, respectively. When matching 50 Ω to 15.6 Ω, 14.3 Ω, 13.3 Ω, or 12.5 Ω, all ratios are constant from 1.5 to at least 30 MHz. The maximum high frequency response varies from 45 MHz with the 1:4 connection down to 30 MHz for the 1:3.2 connection. Tapping at eight or nine turns from terminal 3, for lower impedance ratios, is not recommended.

Right: Thirteen bifilar turns of no. 16 wire on a 1 1/2 in OD, K5 toroid ($\mu = 290$). The top winding (from Figure 7-2) is tapped at 10, 11, and 12 turns from terminal 3. The impedance ratios are 1:3.13, 1:3.41, 1:3.7 and 1:4, respectively. When matching 100 Ω to 29.3 Ω, 27 Ω, or to 25 Ω, the impedance ratios are constant from 1 to at least 30 MHz. The 1:4 ratio, which has the highest frequency response, is constant from 1 to 45 MHz. The 1:3.13 ratio, which has the poorest high frequency response, is constant from 1 to 25 MHz. If the bottom winding in Figure 7-2 is covered with two layers of 3M no. 92 tape, raising the characteristic impedance of the transmission line to 70 Ω, the transformer would have its optimized performance when matching 150 Ω to 48 Ω, 44 Ω, 40.5 Ω, or 37.5 Ω. As before, lower impedance ratios are not recommended.

Chapter 8

Unun Transformer Designs with Impedance Ratios Greater Than 1:4

8.1   Introduction

Transmission line transformers exhibit exceptionally high efficiencies over considerable bandwidths. By connecting several transformers in series, they can provide practical impedance transformation ratios greater than 1:4. This chapter presents techniques for obtaining impedance ratios greater than 1:4:

1. Converting Guanella's transformers, which are basically baluns to unun operation.
2. Adding in series with Guanella baluns higher order winding transformers (e.g., trifilar, quadrifilar) to achieve ratios other than $1:n^2$.
3. Using fractional ratio baluns back to back with Guanella baluns.
4. Applying higher order windings to the Ruthroff type transformer.
5. Tapping these higher order Ruthroff type transformers.
6. Connecting the Ruthroff 1:4 unun in a parallel–series arrangement with a Guanella 1:1 balun.

Guanella's baluns, connected directly as ununs or back-to-back with 1:1 baluns, offer the greatest bandwidths because they add in-phase voltages. On the other hand, Ruthroff's transformers, which add a direct voltage to a delayed voltage or to delayed voltages when higher order windings are used, are much simpler and should find many applications. Further, they can be successfully tapped, yielding a variety of ratios other than 1:4, 1:9, 1:16, and so on.

High impedance transformers are, by far, more difficult to fabricate. The higher impedance translates to higher voltage, which requires more turns to minimize the volts per turn. They require not only higher reactance for sufficient isolation but also windings with higher characteristic impedance. The result is that they are much larger than low impedance transformers, even though their power capabilities are the same. The size of the core (rod or toroid) is related to the magnitude of the characteristic impedance of the transmission line and the number of turns, and not the power level, since very little flux enters the core. Experiments by Sevick showed that characteristic impedances in the range of 150–200 Ω, together with Guanella's modular approach of parallel–series connections, makes possible transformers capable of matching 50 Ω to impedances as high as 1000 Ω with good bandwidth and efficiency. Examples of high-ratio unun transformers are described in the following subsections. Many of these transformers use components described in more detail in Chapters 7 and 9. Of special interest is the fractional ratio balun.

8.2   Guanella Transformers

Guanella showed that by connecting three or more basic building blocks in parallel-series arrangements, impedance ratios of 1:9, 1:16 ... $1:n^2$ are possible with n being a whole number. But these transformers are basically bilateral baluns. Either side can have a grounded terminal, resulting in a step-up or step-down balun. If both sides are grounded, as in the unun case, the low frequency performance can be seriously affected. This is especially true in the 1:4 case when both building blocks are wound on the same core. Extra isolation in the form of a series 1:1 balun or the use of separate cores with more turns is necessary to maintain the same low frequency performance as when operating as a balun. Examples of these were shown with 1:4 ununs. The 1:9 and 1:16 cases, which Sevick found to require separate cores, also improve their low frequency response considerably by using series 1:1 baluns. This is especially true at high impedance levels where the windings, and hence cores, are large. In general, with separate cores for each winding and without a series 1:1 balun, the amateur radio 160 m band is lost when operating the Guanella transformer as a high transformation ratio and high impedance level unun.

Figure 8-1 shows schematics of the high and low frequency models of the Guanella 1:9 transformer. The transformer (Figure 8-1a) is connected in parallel on the left side and in series on the right side, resulting in an output voltage of $3V_{in}$ and hence an impedance ratio of 1:9. Since each transmission line sees one-third of R_L on the right, theory predicts an optimum characteristic impedance of $R_L/3$. In practice, when using low impedance coax cable, the best high frequency results are obtained with characteristic impedances about 90% of theory. Practice has also shown that 1:1 baluns connected in series with Guanella baluns for unun operation need more reactance than the popular 1:1 balun used for isolating Yagi beams and half-wavelength dipoles from coax cables.

Like the Guanella 1:4 transformer, the best low frequency performance occurs when the transformer is connected as a bilateral, 1:9 balun with a floating load (i.e., either terminals 1, 5, 9 or terminal 2 in Figure 8-1a is grounded). But the more interesting case is when the transformer performs as a 1:9 unun (i.e., both terminals 1, 5, 9 and 2 are grounded). In this configuration (with optimized transmission lines), the top transmission line in Figure 8-1a has a gradient of $2V_{in}$ across its input and output terminals, the middle transmission line has V_{in} across its terminals, and the bottom transmission line has zero voltage. Thus, the optimum choice in toroids (or beads) would be ferrites for the top transmission line with a permeability twice that of the middle transmission line. As in the 1:4 case, the bottom transmission line requires no longitudinal reactance and hence no core or beads. Rod core transformers don't enjoy this flexibility in permeability.

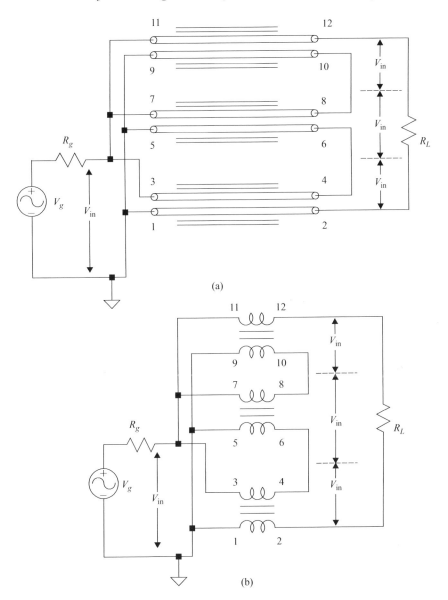

Figure 8-1   *Schematics show mid-band (a) and low frequency (b) circuits for a 1:9 Guanella balun. The load may be center tapped to ground or floating.*

By connecting a 1:1 balun in series at the low impedance side for a step-up unun or at the high impedance side for a step-down unun and removing the ground at terminals 1, 5, and 9 (and using three cores), an improvement of about a factor of two can be realized in the low frequency response.

Another interesting case is when the output voltage is balanced to ground—that is, when grounds are on terminals 1, 5, 9, and 13. In this configuration, the bottom transmission line in Figure 8-1a has a negative gradient of $-3/2V_{in}$, the middle transmission line of $+V_{in}$, and the top transmission line of $+1/2V_{in}$. For best low frequency performance in this case, three cores should be used. Further, the core for the bottom transmission line should have an appropriately higher permeability.

Transformers with ratios of 1:16 and 1:25 can also be designed to perform over wide bandwidths because of the modular nature of Guanella's technique. For example, matching 50 Ω to 800 Ω requires an optimum characteristic impedance of 200 Ω for the transmission lines. This is about the upper limit that can be obtained in power applications with toroids having outside diameters about 2.5 to 3 in.

Matching 50 Ω unbalanced to 600 or 1000 Ω unbalanced requires impedance ratios of 1:12 and 1:20, respectively. These can be obtained by using fractional ratio baluns in series with Guanella baluns. The 50:600 Ω unun can be realized with a 1:1.33 step-up balun in series with a 1:9 balun or a 1.33:1 step-down balun in series with a 1:16 balun. The 50:1000 Ω transformation can be accomplished using a 1.25:1 step-down balun in series with a 1:25 balun.

8.2.1   5.56:50 Ω Ununs

Figure 8-2 shows the schematic of a low impedance, 1:9 unun transformer designed to match 50 Ω coax cable to an unbalanced impedance of 5.56 Ω. A 1:1 Guanella balun is added, in series, on the high impedance side to improve the low frequency performance. Figure 8-3 shows an implementation of the schematic of Figure 8-2. The 1:1 balun has 10 turns of no. 14 wire on a 1 1/2 in OD, 4C4 toroid ($\mu = 125$). One of the wires has two layers of 3M no. 92 tape. This extra separation between the wires results in a characteristic impedance of 50 Ω. Each transmission line of the 1:9 balun has 7 1/2 turns of low impedance cable on a 1/2 in diameter, 2 1/2 in

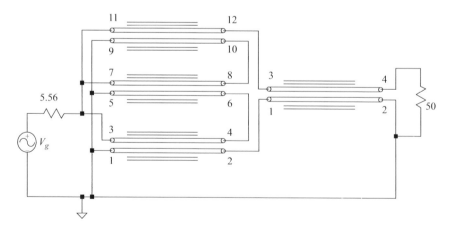

Figure 8-2   *Schematic shows a low impedance unun to match 5.56 Ω to 50 Ω. The 1:1 balun isolates the 1:9 balun so that the output ground connection does not short the 1-2 conductor.*

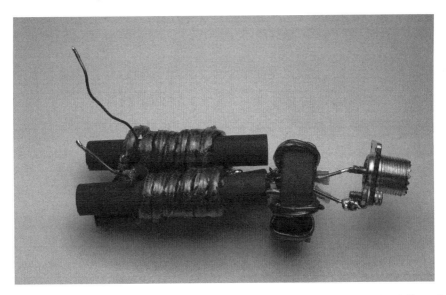

Figure 8-3   Photo shows a 1:9 unun composed of a 1:1 Guanella balun followed by a 1:9 balun. The 1:1 balun isolates the 1:9 balun so the output can be grounded.

long, no. 61 rod ($\mu = 125$). The coax cable uses two layers of 3M no. 92 tape on no. 14 wire for the inner conductor. The outer braid, which is wrapped with 3M no. 92 tape, is from RG-122/U cable or equivalent (1/8 in flat braid can also be used if it is opened up). This transformer has a constant impedance ratio from 3.5 to well over 30 MHz. The highest frequency response occurs at the 6.67:60 Ω level. If no. 12 wire were used instead, the maximum response would occur at the 5.56:50 Ω level. By using a toroid, for the 1:1 balun, with a permeability of 250 or 290, the low frequency response would cover the 160 m band. Further, by using 11 turns of low impedance coax cable on 3 1/2 in long rods, the transformer would cover 160 m through 10 m without the series 1:1 balun. The transformer in Figure 8-2 is conservatively rated at 1 kW continuous power.

This concept can also be extended to a broadband 1:16 unun. In this case, a 1:16 balun would consist of four similar low impedance cables connected in parallel on the left and in series on the right. In this higher ratio case, to better match 3.125 Ω to 50 Ω a thicker wire than no. 14 would have to be used for the inner conductor. A no. 10 wire with 1 1/2 to 2 layers of 3M no. 92 tape is recommended. The outer braid remains the same as in the 1:9 case.

8.2.2   50:300 Ω Ununs

The schematic of a Guanella 50:300 Ω (1:6) unun transformer is presented in Figure 8-4. Several versions of this transformer have been constructed and found to give broadband responses from 1.5 to 45 MHz, depending on the various components employed. They all used the quintufilar, 50:75 Ω unun shown in Figure 8-4

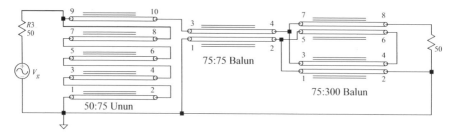

Figure 8-4   Schematic shows a cascade of a Ruthroff transformer and a composite Guanella unun to create a 50:300 Ω unun.

and described in detail in Chapter 7. The 75:75 Ω balun (see Chapter 9) has 12 bifilar turns on a 2 in OD, K5 toroid ($\mu = 290$). Its large reactance, which is some 15 times greater than that of the W2AU balun, was found necessary to provide adequate isolation for the 75:300 Ω Guanella balun. This 1:4 (75:300 Ω) balun, which is also described in Chapter 9, has two bifilar windings, of seven turns each, on a single 2.4 in OD, no. 64 toroid ($\mu = 2.50$). As a balun it performs well from 1.5 to 60 MHz. When used in Figure 8-4, the flat response of the 1:6 unun is from 3.5 to 45 MHz. If this 1:4 balun were replaced with two toroids with 14 bifilar turns each, then the frequency range of the 50:300 Ω unun would be from 1.5 to 45 MHz. A single 2.625 in OD, K5 toroid ($\mu = 290$) with two bifilar windings of eight turns each would also provide the same frequency range. As will be shown in section 8.2.4 and Chapter 9, the 1:1.56 unun and 1:1 balun can be wound on the same toroid, resulting in a fractional ratio balun.

8.2.3   50:450 Ununs

Figure 8-5 is a schematic of a 50:450 Ω unun Guanella transformer. It is simpler, in a way, than the transformer of Figure 8-4, since only two different series transformers are required: a 1:1 balun in series with a 1:9 balun. With optimum design, this transformer can easily handle 1 kW continuous power in a frequency range of 1.5 to over 45 MHz. For the 1:1 balun, a 2 in OD toroid with a permeability of 250 to 290 and with 11 or 12 bifilar turns is recommended (see Chapter 9). The 1:9 (50:450 Ω) Guanella balun is also described in Chapter 9. It uses three 2.625 in OD, K5 toroids ($\mu = 290$), each with 16 bifilar turns of no. 16 wire ($Z_0 = 150$ Ω). Without the 1:1 series balun, the frequency range of the 1:9 unun in Figure 8-5 is 3 to over 45 MHz.

8.2.4   50:600 Ω Ununs

The 50:600 Ω (1:12) unun transformer, covering the HF band (3 to 30 MHz), is one of the most difficult to construct. It requires not only three separate series transformers (unun-balun-balun) but also high characteristic impedance windings to achieve good high frequency performance. The transmission lines, which are then widely spaced, restrict the number of turns on a practical toroid and limit the low frequency response. Sevick was able to design two versions capable of flat responses from 3 to 30 MHz. One used a fractional ratio 1.33:1 step-down balun (two transformers on the same core) in series with a 1:16 step-up balun and a 500 W capability. The other transformer used a fractional ratio 1:1.33 step-up balun in series with a 1:9 step-up balun. It has a 1 kW capability.

The schematic for the first 500 W unun is shown in Figure 8-7. The 1.33:1 step-down balun uses four quintufilar turns of no. 16 wire on a 2 in OD, K5 toroid

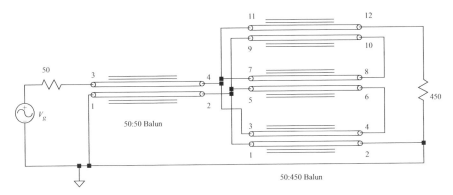

Figure 8-5   *The schematic shows a method to use a 1:1 balun in series with a 1:9 balun to create a 50:450 Ω unun.*

Figure 8-6   *This photo shows a 50:600 Ω unun using Guanella baluns suitable for 1 kW continuous power from 3 to 30 MHz.*

($\mu = 290$). The top winding is tapped one turn from terminal 9. The transformation ratio is 1.4:1. If five quintufilar turns were used and the tap was at two turns from terminal 9, the ratio would be 1.32:1 (but little difference would be noted in performance). The 1:1 balun winding (on the same core) has 11 bifilar turns of no. 16 wire. The 1:16 balun has 13 bifilar turns of no. 20 wire on each of the four 13/4 inch OD, K5 toroids ($\mu = 290$). The spacing of the wires is such as to give a characteristic impedance of 150 Ω.

Figure 8-6 is a photo of the second 50:600 Ω transformer. The schematic for this 1 kW unun is shown in Figure 8-8. The 1:1.33 step-up balun uses three septufilar turns on a 2.4 in OD, K5 toroid ($\mu = 290$). The top winding, 13–14, uses no. 14 wire, and the others use no. 16 wire. The 1:1 balun (on the same toroid) has 11 bifilar turns of no. 16 wire, one of which is covered with two layers of 3M no. 92 tape (for a characteristic impedance of 66.7 Ω). The impedance ratio of this

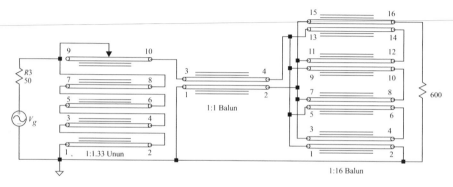

Figure 8-7   This schematic shows a tapped quintufilar transformer in series with a 1:1 balun and a 1:16 balun to create a 50:600 Ω transformer.

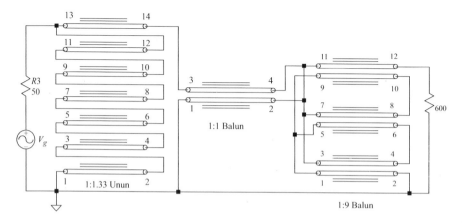

Figure 8-8   Another method of creating a 50:600 Ω uses a septufilar transformer to generate the 1:1.33 ratio before adding a 1:1 balun and 1:9 balun.

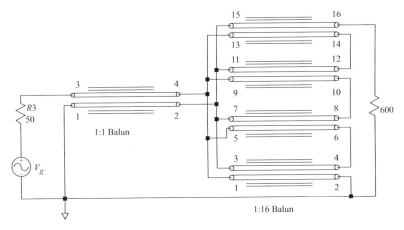

Figure 8-9   This schematic shows a 1:16 unun composed of back-to-back Guanella baluns for operation at the 50:800 Ω level.

fractional ratio balun is 1:1.36 (and is close enough to 1:1.33). The 1:9 balun has 11 bifilar turns of no. 18 wire on each of the three 2.4 in OD, 4C4 toroids ($\mu = 125$). The spacing gives a characteristic impedance of 200 Ω. The wire size is small enough and the radius of the toroid is large enough so that the septufilar winding is no different from the closely wound result using a rod. The difference is that the toroid gives much larger permeability with a closed magnetic path. The same is true for the 1:1 balun, which is wound on the same core.

8.2.5   50:800 Ω Ununs

The schematic for the 50:800 Ω unun Guanella transformer is presented in Figure 8-9. This transformer is also simpler than the 50:600 Ω ununs of Figure 8-7 and Figure 8-8 since it does not require the extra 1:1.33 unun stage. The 50:50 Ω (1:1) balun was illustrated in Figure 8-5. The 50:800 Ω (1:16) balun is a simple extension of the 66.67:600 Ω (1:9) balun (Figure 8-8). In this case, four toroids (with similar windings) are used instead of three.

To match 50 Ω unbalanced to 1000 Ω unbalanced (1:20), which is at the limit of practicality, a suggested configuration is (1) a 1.25:1 step-down unun in series with a 1:1 (40:40 Ω) balun in series with a 1:25 (40:1000 Ω) step-up balun; and (2) the 1:25 balun with five toroids having windings similar to those in Figure 8-8.

8.3   Ruthroff-Type Transformers

Another method of obtaining impedance ratios greater than 1:4 in unun transformers is an extension of Ruthroff's technique of adding a direct voltage to a voltage that has traversed a coiled transmission line (in this case, several coiled transmission lines). Kraus and Allen, using this technique with a shortened third winding, reported a 1:6 ratio [1]. The configuration Sevick used to obtain ratios as high as 1:9 is shown in Figure 8-10. The device has two input ports (A, B) and three output ports (C, D, E). By using B and D, a 1:2.25 ratio is achieved. Using A and D results in a 1:9 ratio. A and C provide a 1:4 ratio. By using A and tapped port E, ratios from 1:4 to 1:9 are possible. By using B and E, ratios from 1:1 to 1:2.25 are obtained.

By connecting an input voltage (V_{in}) to terminal A, the voltage at D is $3V_{in}$ and a 1:9 impedance ratio results. The high frequency response is less than that of the 1:4 ratio and much less than that of the 1:2.25 ratio. This is because the output voltage now consists of a direct voltage, a single delayed voltage, and a double delayed voltage (the top V_{in} in Figure 8-10). The low frequency response is determined in the same manner as that of the 1:4 unun. For example, the reactance of the lower winding in Figure 8-10 should be much greater than the impedance of the signal generator. Conversely, the reactance of the three windings, which will be series aiding, should be much greater than the load impedance. With the 1:9 connection, A to D, two-thirds of the input current (I_{in}) flows in the bottom winding and one-third in the top two windings. As can be seen from the diagram, these currents cancel out the flux in the core, and the high efficiency of a true transmission line transformer can be achieved.

By varying the tap on the upper winding of Figure 8-10, the connection between terminals A and E yields ratios from 1:4 to 1:9. For a rod transformer, the impedance ratio (ρ) becomes

$$\rho = (V_{out}/V_{in})^2 = (2 + l/L) \tag{8-1}$$

where

$l =$ the length from terminal 5
$L =$ the length of the winding

For a toroidal transformer, equation (8-1) is

$$\rho = (2 + n/N)^2 \tag{8-2}$$

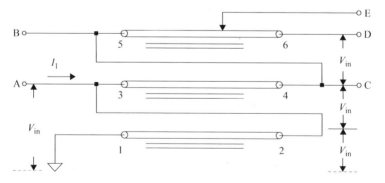

Figure 8-10   Schematic shows a trifilar transformer that can produce a ratio from 1:4 to 1:9 depending on the placement of the tap on the third winding.

where

    n = the number of integral turns from terminal 5
    N = the total number of trifilar turns

Sevick constructed a tapped trifilar transformer using a 1 1/4 in OD, Q1 toroid ($\mu = 125$) consisting of seven trifilar turns of 1/8 in rectangular line insulated with one layer of 3M no. 92 tape. The upper winding (Figure 8-10) is tapped at $n = 3, 4$, and 5 turns from terminal 5. Figure 8-11 shows the performance of this transformer at the various ratios when terminated in a 50 Ω load. With the maximum loss set at 0.4 dB, which is equivalent to a VSWR of 2:1 (i.e., 10% of the power is reflected because of a mismatch), the upper frequency cutoff for all outputs exceeds 30 MHz. The characteristic impedances of the trifilar windings for all ratios, except at 1:4, are optimized for a termination of 50 Ω. The 1:4 ratio response is optimum when $R_L = 30 \ \Omega$.

A tapped trifilar transformer using 1 1/4 in OD, Q1 toroid ($\mu = 125$) is illustrated in Figure 8-12, and Figure 8-13 presents the performance curves of the transformer. This device uses seven trifilar turns of no. 14 wire with taps on the upper winding at $n = 3$ and 5 turns from terminal 5 (only the tap at five turns is connected in the photo). The impedance ratios, using the A port for the input, are 1:4, 1:5.9, 1:7.37, and 1:9. With $R_L = 50 \ \Omega$, Figure 8-13 illustrates that the 0.4 dB cutoff limit is about 15 MHz for ratios greater than 1:4. With the 1:4 ratio, the upper cutoff is 50 MHz. This shows that for a trifilar winding with no. 14 wire and with the top winding floating in the 1:4 configuration, a near optimum condition

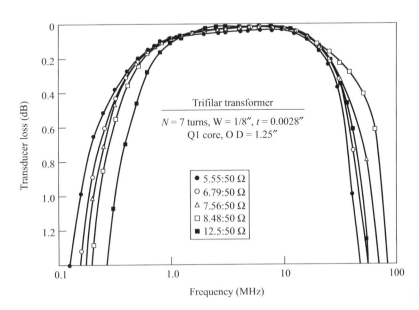

Figure 8-11   *Plots show the response of a tapped trifilar transformer composed of seven turns of 1/8 in rectangular line.*

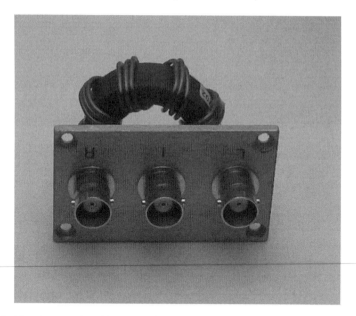

Figure 8-12   A tapped trifilar transformer that is designed to give ratios between 1:4 and 1:9.

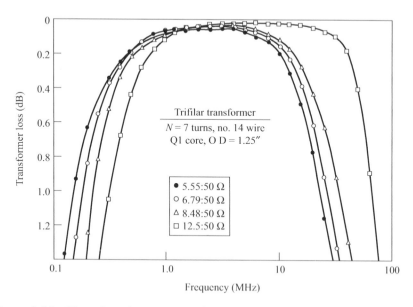

Figure 8-13   Plots show the response of the transformer in Figure 8-12 at various impedance levels. Note the excellent performance at the 12.5:50 Ω level.

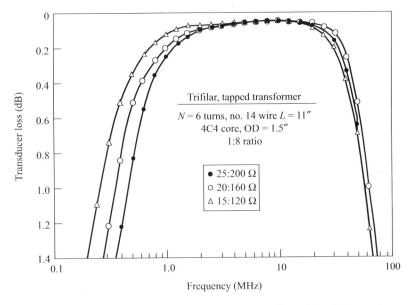

Figure 8-14   Plots show the improved performance of a trifilar 1:8 transformer at higher impedance levels.

exists for a 50 Ω load. When the load impedance is 100 Ω or greater, a higher frequency response is obtained at the higher ratios. Figure 8-14 shows the improved performance at the 1:8 ratio of a trifilar transformer using no. 14 wire and loads of 120, 160, and 200 Ω.

Earlier experiments by Sevick on rod transformers led to similar results. Figure 8-15 gives the performance of a trifilar transformer with no. 14 wire windings, 22 in in length, and tightly wound on a Q1 rod 1/2 in in diameter and 4 in long. The characteristic impedance of the windings using the 1:9 connection was 30 Ω, while the characteristic impedance using the 1:4 connection was only 16 Ω (because of the floating third wire). With the 1:9 connection, better high frequency performance prevails at the 10:90 Ω level. Good high frequency performance is also obtained at the 9:36 Ω level for the 1:4 ratio because of the influence of the floating third wire.

8.3.1   5.56:50 Ω Ununs

Figure 8-13 and Figure 8-15 showed that a 1:9 ratio, at the 5.6:50 Ω impedance level, does not result in good high frequency performance when using no. 14 wire in Figure 8-10. These transformers work better at twice this impedance level. Figure 8-15 also demonstrates that 12 trifilar turns on a 1/2 in diameter rod are more than necessary to include 1.5 MHz at the low frequency end. To improve the high frequency performance of these transformers at the 5.56:50 Ω level, Figure 8-10 was rearranged to produce Figure 8-16. By transposing the bottom

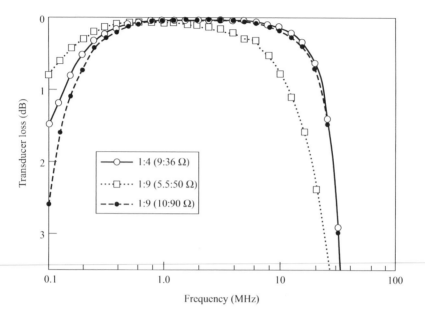

Figure 8-15   Plots show the response of a trifilar transformer wound on a ½ inch rod for ratios of 1:4 and 1:9.

Figure 8-16   Rearranging Figure 8-10 lowers the effective impedance of the transmission line as well as the optimum impedance level.

winding in Figure 8-10 to the middle winding, the characteristic impedance was lowered considerably. By selecting the optimum number of turns and using the circuit in Figure 8-16, transformers can be made to adequately cover the 1.5 to 30 MHz range, even though their optimum performances occur at nearly twice this impedance level. At these higher impedance levels, the high frequency performance extends well beyond 45 MHz.

Figure 8-17 shows two transformers from Figure 8-16 with the following parameters:

Left: Five trifilar turns of no. 14 wire on a 1 1/2 in OD, 4C4 toroid ($\mu = 125$) with 1 kW continuous power.

Figure 8-17   Photo shows two 1:9 ratio ununs using the schematic in Figure 8-16 to obtain operation from 1.5 to 30 MHz at the 5.56:50 Ω level.

Right: Seven trifilar turns on a 1/2 in diameter, no. 61 rod ($\mu = 125$). The middle winding in this transformer is no. 12 wire. The other two are no. 14 wire. The power rating is 1 kW continuous power.

If both transformers in Figure 8-17 used no. 12 wire for the middle winding in Figure 8-16 and no. 14 wire for the outer two, they would all have the same power rating of 1 kW continuous power. As has been noted before, since so little flux enters the core, the power rating is actually determined by the ability of the windings to handle the current and not by the size of the core.

8.3.2   50:450 Ω Ununs

As mentioned already, the trifilar transformers in Figure 8-13 and Figure 8-15 (both use the schematic of Figure 8-10), are best suited to match at impedance levels considerably higher than 5.56:50 Ω. But going up to a much higher impedance level, like trying to match 50 Ω to 450 Ω, presents a most difficult task for the Ruthroff-type transformer. Not only are many more turns required at this impedance level, but also characteristic impedances in excess of 200 Ω only compound the problem. Nevertheless, transformers matching 50 Ω to 450 Ω, having a relatively constant 1:9 impedance ratio from 1.5 to 10 MHz, are rather easy to construct and should have some practical uses. One example is a matching transformer for a Beverage antenna. Figure 8-18 presents this type of transformer together with a 1 kW power transformer with the following parameters:

Left: Twelve trifilar turns on a 2.4 in OD, 4C4 toroid ($\mu = 125$). The bottom winding uses no. 14 wire and the other two no. 16 wire. All three windings are covered with Teflon tubing (wall thickness about 17 mil). As with the

Figure 8-18   Transformers with high impedance trifilar windings are suitable for use from 1.5 to 10 MHz. They match 50:450 Ω with 1:9 ratio.

smaller transformer, the 1:9 ratio is constant from 1.5 to 10 MHz. The power rating is at least 1 kW continuous power.

Right: Thirteen trifilar turns of no. 22 hook-up wire on a 1 1/4 in OD, Q1 toroid ($\mu = 125$). Ratios 1:4 and 1:9 are available and are constant when matching 50 Ω to 200 Ω or to 450 Ω from 1.5 to 10 MHz. The power rating is 100 W continuous power. This transformer is intended for use with a Beverage antenna.

8.3.3   3.125:50 Ω Ununs

The final part of this section presents two Ruthroff-type transformers that produce 1:16 ratios at low impedances with only single cores: two coax cable 1:4 ununs in series (Figure 8-19a); and quadrifilar windings (Figure 8-19b). Figure 8-20 is a photograph of the transformers. It is important that the windings are designed for series aiding for low frequency operation. The parameters for the transformers are as follows:

Left: The low impedance unun (Figure 8-19a, left) has four turns of coax cable. The inner conductor is no. 10 wire with about 1 1/2 layers of 3M no. 92 tape (because it is wrapped edgewise). The outer braid, from RG-122/U, is wrapped with 3M no. 92 tape. The characteristic impedance is 9 Ω. The higher impedance unun (Figure 8-19a, right) has seven turns of coax cable. The inner conductor, of no. 16 wire, has two layers of 3M no. 92 tape. The outer braid, from RG-122/U, is also wrapped with 3M no. 92 tape. The characteristic impedance is 14 Ω. The toroid is a 1 1/2 in OD, 4C4

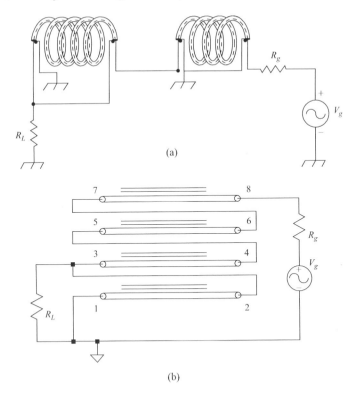

(a)

(b)

Figure 8-19   Schematic representations show connections for a coax or bifilar 1:16 transformer using series connected 1:4 transformers.

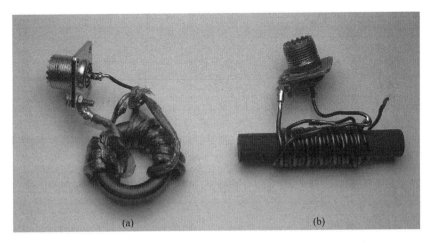

Figure 8-20   Transformers implement a 1:16 ratio transformer on the same core. (a) Low impedance coax cable on a toroid. (b) Quadrifilar 1:16 transformer on a rod.

core ($\mu = 125$). The transformer covers 1.5 to 30 MHz when matching 3.125 Ω to 50 Ω. The highest frequency response occurs when matching 3.875 Ω to 62 Ω. The power rating is 1 kW continuous power.

Right: Seven quadrifilar turns on a 1/2 in diameter, no. 61 rod ($\mu = 125$). The bottom winding in Figure 8-20b uses no. 14 wire, and the other three windings use no. 16 wire. At the 3.125:50 Ω level, the impedance ratio is constant from 1.5 to 10 MHz. The highest frequency response occurs at the 6.25:100 Ω level, and the impedance ratio is constant up to 25 MHz. The power rating is 500 W continuous power.

8.4   Ruthroff-Guanella Transformers

Figure 8-21 presents another technique for a 1:9 unun transformer. This method is an extension of Guanella's method on connecting transformers in a parallel–series arrangement. In Figure 8-21, the top two windings form a 1:1 balun transformer, and the bottom two windings are connected to form a 1:4 Ruthroff unun. The transformers are in parallel on the left side and in series on the right. Since the outputs of the transformers are isolated from the inputs in the passband because of the reactance of the coiled windings, the output voltage is $V_{out} = 3V_{in}$. This results in a 1:9 impedance ratio when the output is taken from terminal B to ground. If the output is taken from terminal C to ground, a 1:4 ratio is obtained.

Sevick created a transformer (Figure 8-21) using two 1 3/4 in OD, K5 toroids ($\mu = 290$). The top two windings were 16 bifilar turns of no. 16 wire. The bottom two windings were eight bifilar turns of no. 16 wire. The useful impedance level range is 13.33:120 to 20:180 Ω, where the impedance ratio is constant from 1.5 to 30 MHz. When matching 15.56:140 Ω, which is the optimum impedance level, the ratio is constant from 1.5 to 45 MHz. The power rating is 1 kW continuous power.

Sevick compared the combination Ruthroff-Guanella transformer to a Ruthroff-type transformer. The latter has eight trifilar turns of no. 16 wire on a 1 3/4 in OD,

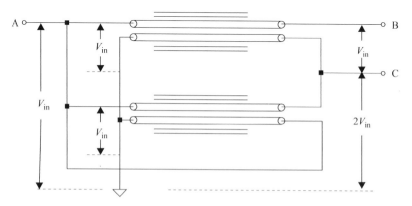

Figure 8-21   Schematic shows a combination Guanella-Ruthroff transformer to achieve a 1:4 or 1:9 ratio.

K5 toroid ($\mu = 290$) and is connected as shown in Figure 8-10. Compared with the combination transformer, the Ruthroff-type transformer's useful frequency range is about one-half (i.e., only 1.5 to 15 MHz). There is little doubt that the Ruthroff-Guanella transformer is superior to the Ruthroff-type transformer in this range of impedance levels.

From the results on these and other Ruthroff-Guanella transformers, the following comments are offered:

1. The Ruthroff-Guanella transformer requires two cores for best operation.
2. When using low impedance coax cables (e.g., when matching 5.56 Ω to 50 Ω), satisfactory operation was obtained from 1.5 to only 25 MHz. Beyond 25 MHz, serious resonances occurred.
3. The Ruthroff-Guanella transformer was difficult to work with at impedances greater than 30.
4. The top two bifilar windings determine, in large measure, the low frequency response. It is recommended that they have twice the number of bifilar turns as the bottom two windings. With fewer turns than this, the high frequency/performance improves, but at the expense of the low frequency performance.
5. The regular Guanella transformer with its parallel–series connection of transmission lines is by far the best transformer.

8.5   Coax Cable Transformers—Ruthroff Type

As has been illustrated throughout this book, many wire versions of transmission line transformers can be converted to coax cable transformers. The resulting advantages are higher current and voltage capabilities and less parasitic coupling between adjacent turns. Further, they also lend themselves quite readily to multi-port operation: that is, they can possess more than one broadband impedance ratio. They can also be tapped, yielding fractional ratios. Figure 8-22 shows a schematic diagram for a trifilar, configuration yielding ratios of 1:2.25 with terminals A and B, 1:4 with terminals B and C, and 1:9 with terminals A and C. Sevick created a trifilar, coax cable transformer using two 15 in sections of RG-58/U on a 2.62 in K5 toroid. The outer covering of the coax was removed for ease of winding. As shown

Trifilar coaxial transformer

Figure 8-22   Schematic shows the connections for coax trifilar transformer.

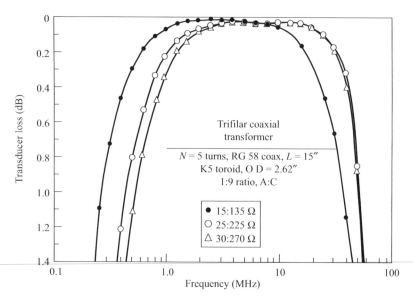

*Figure 8-23   Plots show the response of the coax trifilar transformer when
connected for a 1:9 ratio. The plots show performance versus
impedance ratio and frequency.*

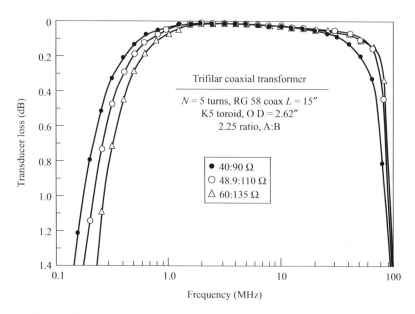

*Figure 8-24   Plots show the response of the coax trifilar transformer when
connected for a 1:2.25 ratio.*

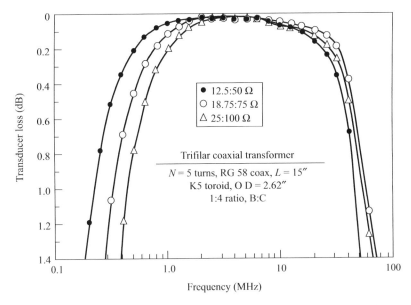

Figure 8-25   Plots show the response of the coax trifilar transformer when connected for a 1:4 ratio.

in Figure 8-22, the outer braids are at the same potential and therefore can be in contact with each other. Figure 8-23 presents the performance with the 1:9 ratio for three different levels of output impedance: 135, 225, and 270 Ω. At the high frequency end, the 0.4 dB points are at 40 MHz for the 225 and 270 Ω loads. Surprisingly, between 2 and 25 MHz, the losses are less than 0.1 dB at these impedance levels. This is considerably less than wire transformers using this rather high permeability K5 material ($\mu = 290$). The 1:2.25 performance in Figure 8-24 is equally interesting. The upper cutoff frequencies approach 80 MHz with 110 and 135 Ω loads. Figure 8-25 gives the performance for the 1:4 connection. These results demonstrate the influence of the trifilar connection. Optimum performance now occurs at the 18.75:75 Ω level instead of the 25:100 Ω level. A greater slope in the loss with frequency than for the other two ratios is evident in Figure 8-25. This slope is typical of wire transformers using K5 material.

Reference

[1]   Krauss, H. L., and C. W. Allen, "Designing Toroidal Transformers to Optimize Wideband Performance," *Electronics*, Aug. 16, 1973.

Chapter 9

Baluns

9.1   Introduction

This chapter covers baluns, the subset of transmission line transformers of most interest to antenna builders since most antenna structures have symmetrical feed points. We will look at both Ruthroff- and Guanella-style baluns.

9.2   The 1:1 Balun

The 1:1 balun is well known to radio amateurs and antenna professionals since it is widely used to match coax cables to dipole antennas and to Yagi beams that incorporate matching networks which raise the input impedance to that of the cable. The purpose of the balun is to minimize RF currents on the outer shield of the coax cable which would otherwise distort radiation patterns (particularly the front-to-back ratio of Yagi beams) and also cause problems because of RF penetration into the operator location. The balun accomplishes this by suppressing any induced antenna current on the outer coax conductor due to antenna asymmetry. In other words, a 1:1 balun is intended to operate as a common mode choke whose job is to suppress common mode energy (energy flowing on the outside of the coax shield) and allow differential mode energy to flow unimpeded. Many successful forms of the 1:1 balun have been used. They include: (a) the bazooka which uses ¼ λ decoupling stubs, (b) 10 turns of the coax line with a diameter of 6 to 8 in, (c) ferrite beads over the coax line, and (d) ferrite core or air core Ruthroff designs.

Figure 9-1 presents what is probably the most popular form of the 1:1 balun, the Ruthroff design, and illustrates the toroidal and rod versions. The third winding of the toroidal transformer, shown as an inductor rather than a loaded transmission line, in Sevick's experience was usually wound on its own part of the toroid. This has the effect of turning the balun from a transmission line transformer to a hybrid using both transmission line and magnetic transformer qualities. A true transmission line transformer would be obtained by using a trifilar winding on the toroid just as we do on the rod. The low frequency model for the 1:1 baluns in Figure 9-1 is shown in Figure 9-2.

Ruthroff originally considered the third wire, winding 5–6, as necessary to complete the path for the magnetizing current. In discussions between Sevick and his colleagues, including Ruthroff, they agreed that the third wire is not necessary

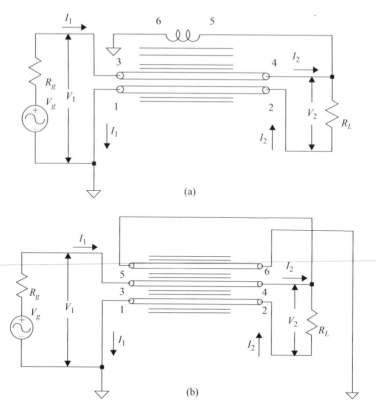

Figure 9-1 *Schematics show Ruthroff versions of baluns: (a) Toroid version of a balun, where an isolated third winding is used. (b) Rod version using a trifilar winding, where all three wires are closely coupled.*

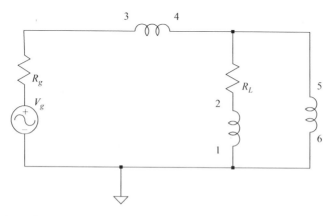

Figure 9-2 *Schematic shows the low frequency model of the Ruthroff 1:1 balun.*

in the performance of the Ruthroff 1:1 balun in antenna applications. When the reactance of the windings is much greater than R_L (at the lowest frequency of interest), then only transmission line currents flow and there is no magnetizing current.

Both Sevick and Ruthroff place the third wire (winding 5–6) at a $+V_{in}/2$ potential. This is correct only for the case where all three windings are closely coupled, as in the case of the rod balun or when a toroid is wound with all three windings closely coupled. When all three wires are closely coupled, the outer wires carry only one-half of the current and the center wire carries the full current of the load.

As Sevick observed, if the energy flow is purely by transmission line current, no net field is transferred into the core. This means there is no voltage across the isolated third winding of the toroid design. If there is any imbalance in transmission line mode, a net field will occur in the core and the 5–6 winding will generate a voltage. The voltage at terminals 4 and 5 is a function of the level of imbalance with the limit reached when the voltage is $+V_{in}/2$. Notice that the connection is opposite of what we would use for a 1:9 transformer. The voltage at terminal 5 is in phase with the voltage (V_{in}) and tends to hold terminals 4 and 5 with no potential to ground at balance. Even though the potential at terminal 5 is the same as ground, the load is totally isolated from the ground at the input of the transformer. Any current that flows in or out of terminal 5 will force the voltage to be different from ground.

Once the coupled windings no longer operate as a transmission line, the system falls apart quickly. For the transformer to actually be a transformer, the voltage is identical for all three windings and the current is identical for all three windings (assuming all three wires are identical). The requirement for current to be identical cannot be supported in magnetic transformer operation because the current in winding 1–2 will always be the sum of winding 3–4 and 5–6, which are in phase. It is possible to have a balanced system as frequency decreases, but the phase of the respective currents must change due to unbalanced parasitic elements such as capacitance and leakage inductance. Mesh analysis of Figure 9-2 shows that the system cannot operate as a magnetic transformer at low frequencies.

Sevick and Ruthroff are correct for the cases where the windings act like three coupled wires and the energy flow is by transmission line mode. The 3–4 center wire carries twice the current of the other two wires (5–6 and 1–2), so the voltage at the load with respect to ground is $+V_{in}/2$ at terminals 4, 5 and $-V_{in}/2$ at terminal 2.

On the other hand, the Guanella 1:1 balun in Figure 9-3, which is nothing more than a coiled bifilar winding, does not have the negative effects of the three-wire Ruthroff balun. At very low frequencies, the Guanella balun reduces to a tightly coupled autotransformer. In the passband where the load is isolated from the input of the 1:1 balun (due to coiling), the center of R_L can be connected at ground or common potential without disrupting the balance of the circuit. If the load is not connected to common or ground, it can float to any voltage induced by common mode effects, again without affecting the balance of the circuit at the load.

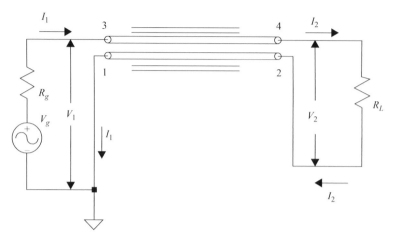

Figure 9-3   The schematic shows the operation of a Guanella 1:1 balun. The center of R_L can be connected to ground without disrupting the balance of the circuit.

The input impedances of the 1:1 baluns in Figure 9-1 and Figure 9-3 are the same as that of a terminated transmission line. Assuming sufficient choking such that only transmission line currents flow and neglecting parasitic elements between adjacent bifilar turns, the input impedances are

$$Z_{in} = Z_0(Z_L + jZ_0\tan \beta l)/(Z_0 + jZ_L\tan \beta l) \tag{9-1}$$

where

Z_0 = the characteristic impedance
Z_L = the load impedance
l = the length of the transmission line
$\beta = 2\pi/\lambda$, where λ = the effective wavelength in the transmission line

Equation (9-1) shows that the input impedance can be complex, except when $Z_0 = Z_L$, and is periodic with the variation of βl – period being π or $l = \lambda/2$. For short transmission lines (i.e., $l < \lambda/4$), the impedance is less than Z_L if Z_L is greater than Z_0 and greater than Z_L if Z_L is less than Z_0. In other words, the transformation ratio is greater than 1:1 if Z_L is less than Z_0 and less than 1:1 (such as 0.5:1) if Z_L is greater than Z_0. This variation in the transformation ratio becomes apparent when the length of the transmission line becomes greater than 0.1 λ.

The characteristic impedance of the 1:1 balun is assumed to be the same as that of the coax cable that is connected to its terminals. This is true with the Guanella balun using no. 14 or 16 wire with very little spacing between the wires and ample spacing (at least one diameter spacing) between adjacent bifilar turns. However, when extra insulation such as Teflon tubing is employed, the characteristic impedance can become two or three times greater than that of the coax cable, and the input impedance can differ widely from that of 50 Ω cable, even at reasonably low frequencies.

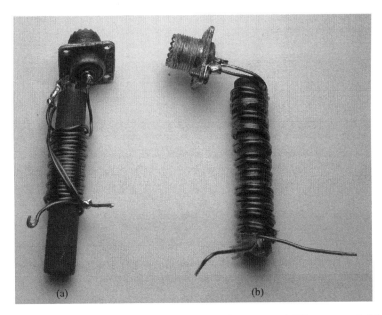

Figure 9-4 *(a) Ruthroff trifilar rod balun similar to a W2AU commercial balun.*
(b) Rod version of a Guanella balun.

Figure 9-4 shows two 1:1 baluns, capable of handling 1 kW continuous power, using 1/2 in diameter rods with permeabilities of 125, with the following parameters:

Left: A Ruthroff design balun using eight trifilar turns of no. 16 wire on a three in long, 1/2 in diameter rod equivalent to a commercial W2AU balun. The characteristic impedance is 43 Ω. The third wire (Figure 9-1b, winding 5–6) is placed between the other two windings. Without it, the characteristic impedance of a tightly wound bifilar winding of no. 16 would be 25 Ω. This balun has been widely used on triband (10, 15, and 20 m) Yagi beams. At much lower frequencies, the performance becomes marginal. It is recommended that this balun not be used below 3.5 MHz.

Right: A Guanella balun using 12 bifilar turns of no. 16 wire on a 3 in long rod. One of the windings is covered with two layers of 3M no. 92 tape. This added insulation, together with the wire diameter spacing between adjacent bifilar turns, raises the characteristic impedance to 50 Ω. The low frequency response of this balun is the same as that of the W2AU balun. If a polyimide coated wire such as ML or H Imideze wire were used, together with one wire having the two layers of 3M no. 92 tape, the breakdown of this transformer would rival that of coax cable. As with the Ruthroff design on the left, this is a satisfactory balun for triband Yagi beams. This balun is capable of operating from 1.7 to 30 MHz. If operation is limited to the 40, 80, and 160 m bands, then 14 bifilar turns are recommended. This would allow more margin on 160 m.

9.2.1   Rod versus Toroidal Baluns

Since the 1:1 baluns in Figure 9-4 appear to satisfy most amateur radio needs and are the simplest and most inexpensive baluns to construct, why use toroids for the cores? First, because the toroidal transformer has a closed magnetic path and the permeability plays a direct role in the reactance of the coiled windings (rod transformers are independent of permeability), much greater margins can be obtained at both the low and high frequency ends of operation. This allows fewer turns to be used to obtain the desired margins. The low frequency response with the toroid is better than a rod transformer by a factor of 2.5, which is also the ratio of their inductances. Further, higher permeabilities, on the order of 250–300, can still be used with good efficiencies. Therefore, the overall improvement is greater by a factor of at least 5. Also, the toroid lends itself more readily to the use of thicker wires (especially no. 10 and no. 12) and coax cables, which allows higher power levels. Since fewer turns are needed with toroids, the spacing between the bifilar turns or the coax cable turns can be increased, which lowers the parasitic coupling and increases the high frequency response. In addition, if 1:1 baluns are required in the 75–200 Ω range, then the toroidal core is, without doubt, the best choice. The rod core would greatly restrict the useful bandwidth. And finally, many symmetrical forms of the combined balun/unun transformer as well as Guanella baluns employed in unun operation use 1:1 baluns in series with 1:4 (and higher) ratio baluns. The extra isolation (from ground) offered by these 1:1 baluns is necessary to preserve the low frequency response. The best isolation is obtained with toroidal core 1:1 baluns.

Figure 9-5 presents two 1:1 Guanella baluns using toroids with the following parameters:

Left: 11 bifilar turns of no. 12 H Imideze wire on a 2.4 in OD, 52 toroid ($\mu = 250$). Each wire has two layers of 3M no. 92 tape, resulting in a characteristic impedance of 50 Ω. The power and voltage capabilities are approximately 5 kW continuous power and 5 kV, respectively. The useful bandwidth extends from 1 to well over 50 MHz. Compared with the W2AU balun in Figure 9-4, an engineering estimate by Sevick indicates that this transformer is 5 to 10 times better on power, voltage, and bandwidth. It can also be duplicated with no. 14 H Imideze or ML wire, with only one winding having two layers of 3M no. 92 tape. The power and voltage capability would then be reduced by a factor of only about 2.

Right: This is an example of a 75 Ω, 1:1 Guanella balun. It has 10 bifilar turns of no. 14 wire on a 2.4 inch OD, 52 toroid ($\mu = 250$). One wire is covered with 17 mil wall Teflon tubing to increase the characteristic impedance to 75 Ω. The response is useful from 1.5 to over 50 MHz. The power rating is about 2 kW continuous power. The voltage breakdown with H Imideze or ML wire is in excess of 2 kV. Baluns for 1:1 operation can be constructed with characteristic impedances up to 200 Ω.

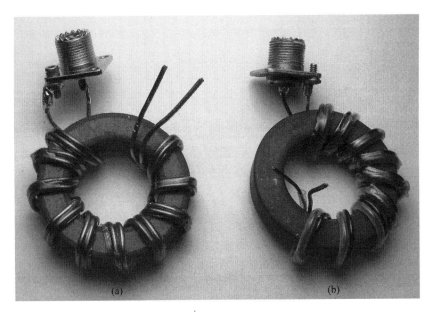

Figure 9-5   (a) 50 Ω 1:1balun using no. 12 wire. (b) 75 Ω 1:1 balun using no. 14 wire and PTFE tubing on one wire.

9.2.2   Bifilar versus Trifilar Baluns

Most Yagi beam antennas employ a shunt feed system to raise the input impedance to values near that of coax cables. In many cases this results in the center of the driven element being at a common potential. Physically and mathematically, this can be represented by the input impedance of the Yagi beam being grounded at its center point. This beam antenna, which is balanced to ground, is then matched to an unbalanced coax cable by a 1:1 balun. The question then is: how do the Ruthroff and Guanella baluns work in this case? The trifilar Ruthroff balun performs satisfactorily as witnessed by its significant use. What about the bifilar Guanella balun? To understand the problems involved with both of these baluns, the low frequency models in Figure 9-6 were studied, together with measurements obtained on baluns with "floating" loads (like that of dipoles) and grounded, center tapped loads (like that of Yagi beams). The numbers on the ends of the windings in the figure were taken from their high frequency models (i.e., Figure 9-3 for the Guanella model and Figure 9-1 for the Ruthroff model).

The low frequency models in Figure 9-6 assume that the transmission lines formed by windings 1–2 and 3–4 (in both cases) are completely decoupled as far as transmission line operation is concerned and that the windings are coupled only by flux linkages. There is virtually no energy being transmitted by a transmission line mode in this case. This condition arises when R_L is much greater than the reactance of the individual windings. As the figure outlines, if the frequency is lowered to a point where the reactance approaches zero, then the impedance of the Guanella

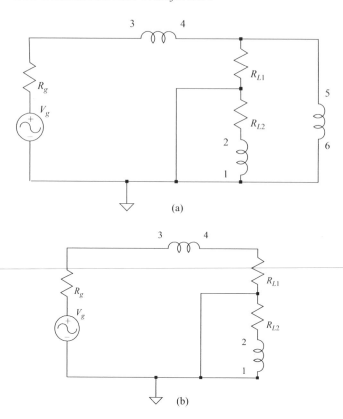

Figure 9-6 Schematics show the low frequency equivalent circuits for balanced center tapped load connected to ground: (a) Three-wire Ruthroff balun. (b) Two-wire Guanella balun.

balun approaches $R_L/2$ and the impedance of the Ruthroff balun approaches zero. But if the reactance of the windings is much greater than R_L (at least 10 times greater), then energy is mainly transmitted to the load by a transmission line mode and the grounding of the center of the load becomes unimportant. To prove this point, baluns from Figure 9-4 and Figure 9-5 were measured for their low frequency responses with the center tap of R_L grounded (where the problem could arise) with the following results:

1. The rod (W2AU) balun on the left in Figure 9-4: The low frequency response at 3.5 MHz is the same whether the center of R_L is grounded or not. The shunting effect of winding 5–6 dominates at the low end. In any case, grounded or not, it is recommended that this balun be used only above 3.5 MHz.
2. The rod Guanella balun on the right in Figure 9-4: The acceptable low frequency response is 3.5 MHz with the center of R_L grounded. Without the center of R_L being grounded, the low frequency limit is 1.7 MHz.
3. The toroidal Guanella balun on the left in Figure 9-5: This balun, which has 11 bifilar turns on a 2.4 in OD, 52 toroid ($\mu = 250$), has a much greater

reactance in its windings than the original rod transformers. Measurements showed that the acceptable low frequency response with the center of R_L grounded is as low as 1 MHz. This is the balun to use if a shunt fed Yagi beam were to be used on 160 m!

9.2.3   Air Core versus Ferrite Core Baluns

When ferrite core transmission line transformers are designed and used properly (i.e., when the reactance of the coiled windings, at the lowest frequency of interest, is at least 10 times greater than the effective termination of the transmission lines), the currents that flow are mainly transmission line currents. This mode of operation leads to the very wide bandwidth and the exceptional efficiency (virtually no core loss) that is achievable with these transformers. The core losses and the high frequency responses of 1:1 baluns are primarily determined by the properties of the coiled transmission lines performing as RF chokes.

Air core baluns do eliminate the core problems when the wrong ferrite is used or when the reactance of the coiled windings is insufficient to suppress the longitudinal currents that create core flux. These currents are the conventional transformer currents and the induced antenna currents on the feed line. With good antenna and feed line symmetry and the proper choice of feed line length (the length to ground of the feed line and one-half of the dipole should not be an odd multiple of $\lambda/4$), the induced currents can become insignificant. But the problem with air core baluns is the inordinate number of turns required to achieve reactance comparable to ferrite core baluns. A winding on a 4 in long ferrite rod has more than 10 times the reactance of a similar air core winding. When comparing the reactance of an air core winding with a similar one on a toroid, the difference is even more dramatic: 27 times more reactance with a toroid of permeability 125. With a toroid having a permeability of 290, this amounts to a difference of 62 times. In other words, to equal the reactance provided by a ferrite toroid, the air core balun would require seven to eight times more turns. Therefore, to approach the isolation properties of the balun shown on the left in Figure 9-5, an air core balun would require 80 bifilar turns. But this is not the whole story. The high frequency properties of an RF choke with 80 turns must be considered. Obviously, its self-resonance (which is the limiting factor at the high frequency end) is a much lower frequency than that of the 11-turn balun in Figure 9-5. In summary, the air core balun does eliminate potential core problems associated with ferrite core baluns, but at the expense of bandwidth.

9.3   The 1:4 Balun

The 1:4 balun, although not as popular as the 1:1 balun, has found considerable use in antenna applications. These include matching folded dipoles to coax cables and matching balanced feed lines to unbalanced networks in antenna tuners. Like the 1:1 baluns in section 9.2, the 1:4 balun also has two forms: Ruthroff and Guanella. Their high frequency models are shown in Figure 9-7, and Figure 9-8 presents their low frequency models. Unlike the 1:1 baluns, both forms of the 1:4 balun have a

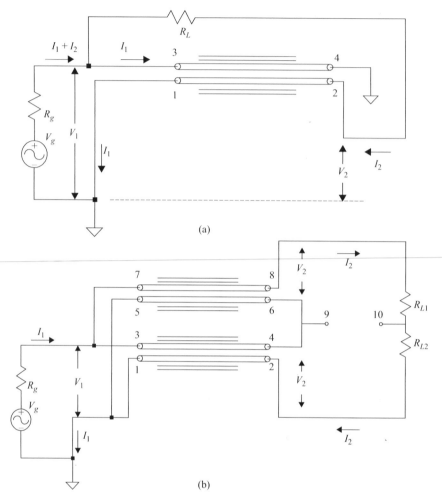

Figure 9-7   (a) High frequency schematic of a Ruthroff 1:4 balun. (b) High frequency schematic of a Guanella 1:4 balun.

considerable difference in their performance, depending on whether the load (R_L) is floating or connected to a common potential at its midpoint.

The analyses of these two conditions are as follows.

(A) R_L Floating

The Ruthroff balun (Figure 9-7a) works on the principle that the left side of R_L is at $+V_{in}$ due to the direct connection to terminal 3 and the right side is at $-V_{in}$ (for a matched transmission line) because of the negative gradient across both windings (Chapter 1, section 1.2). Then $V_{out} = 2V_{in}$ and the impedance ratio is 1:4. Further, the output is balanced to ground. If the reactance of the windings is much greater

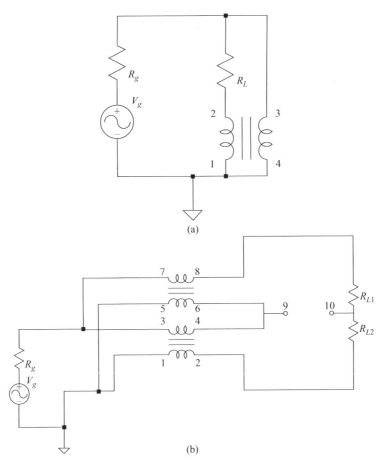

Figure 9-8 *(a) Low frequency equivalent of a Ruthroff 1:4 balun. (b) Low frequency equivalent of a Guanella 1:4 balun.*

than R_L at the frequency of interest, the currents that flow are only transmission line currents. The high frequency performance is the same as that of the Ruthroff unun (Chapter 6). On the other hand, the Guanella 1:4 balun (Figure 9-7b) obtains a doubling of the input voltage (V_{in}) by simply adding the outputs of two transmission lines. These coiled transmission lines are connected in parallel at the low impedance side and in series at the high impedance side. Three important distinctions are noted in these two approaches, when using a floating load:

1. The Ruthroff balun works only in one direction. The high impedance side (Figure 9-7a, right) is always the balanced side. The Guanella balun, on the other hand, is bilateral. It can work as well in either direction, depending on which terminal (1, 5, or 2 in Figure 9-7b) is grounded. Therefore, a Guanella balun can easily be designed to match a 50 Ω coax cable to a 12.5 Ω balanced load.

2. The high frequency response of the Ruthroff balun is considerably less than that of the Guanella balun since it adds a delayed voltage to a direct voltage. At the frequency where the delay is 180°, the output of the Ruthroff balun is zero. Since the Guanella balun adds two voltages of equal phases, the upper frequency limit is mainly determined by the parasitic elements because of the coiling of the transmission lines. If the transmission lines are effectively terminated in their characteristic impedances ($R_L/2 = Z_0$) and parasitic elements are minimized, then the impedance ratio of the Guanella balun is essentially frequency independent.

3. The Guanella concept is a modular approach that can be extended to yield higher impedance ratios. Three transmission lines can easily be connected in a parallel–series arrangement, resulting in a broadband 1:9 balun; four transmission lines would result in a 1:16 balun; and so on. With a practical limit of about 200 Ω, which is obtainable for the characteristic impedance of a coiled transmission line, efficient, broadband baluns matching 40–1000 Ω are possible. The Ruthroff balun cannot possibly compete.

Another interesting comparison is seen in their low frequency models (Figure 9-8). If the total number of turns in the two coils of the Ruthroff balun is the same as the total in the four coils of the Guanella balun, and if a single toroid is used (making sure the windings in the Guanella balun are in series-aiding), then the low frequency responses are identical. If the two transmission lines of the Guanella balun are wound on separate toroids and the total number of turns on each toroid equals the total number on the Ruthroff balun, then the low frequency response of the Guanella balun is better by a factor of two. In either case—the single- or two-toroid—the Guanella balun has a much higher frequency response since it adds in-phase voltages.

(B) R_L Grounded at Midpoint

A different condition arises when the load (R_L) is center tapped to ground. The low frequency response of the Ruthroff balun is essentially unchanged. But the high frequency response, unexpectedly, takes on the nature of a Guanella balun; that is, measurements show that the high frequency response is vastly improved, indicating that two in-phase voltages are now being summed. This could be of interest in designing combination balun/unun transformers or 1:4 baluns with loads center tapped to ground. On the other hand, the Guanella balun is quite seriously affected in its low frequency response when a single core is used. Figure 9-8 shows that winding 1–2 has $R_L/2$ directly across its terminals. Thus, the reactance of winding 1–2, alone, should be much greater than $R_L/2$. This "loading" of winding 1–2 is also reflected into the other three windings because of the tight magnetic coupling at low frequencies. In this case, a series 1:1 balun is necessary to restore the low frequency response to the floating R_L condition. If two separate toroids are used with the Guanella 1:4 balun, then center tapping R_L to ground has only a small effect.

The following examples of 1:4 baluns are grouped according to impedance levels. Comparisons are shown between Ruthroff and Guanella baluns as well as between ferrite and powdered iron core baluns. The latter comparison is particularly directed toward the use of 1:4 baluns in antenna tuners. This does not answer all of the questions related to antenna tuners, since the total solution involves the design of the L-C network and the length and character of the feed line. The complete design of antenna tuners is beyond the scope of this book. The examples presented in this section on low impedance baluns use only Guanella's approach. Since his baluns are bilateral, excellent 1:4 baluns matching 50 Ω coax to 12.5 Ω are readily designed and should find use in matching coax directly to Yagi beam antennas without delta matches, hairpins, and so on. Other baluns having ratios less than 1:4, which can be used for Yagi beams of various element spacing as well as for quad antennas, are described later in the chapter.

9.3.1   *50:200 Ω Baluns*

Figure 9-9 shows two 1:4 baluns designed to match to floating balanced loads in the range of 150 to 300 Ω. Their best high frequency responses occur when matching 50 Ω coax to 200 Ω. The characteristic impedance of their windings is about 100 Ω with the following parameters:

Left: This Ruthroff balun has 15 bifilar turns of no. 14 wire on a 2.4 in OD, 4C4 toroid ($\mu = 125$). Each wire is covered with 17 mil wall Teflon tubing. The reactance, due to 15 turns, is sufficient to allow efficient operation down to 1.5 MHz. The coax connector is on the low impedance side.

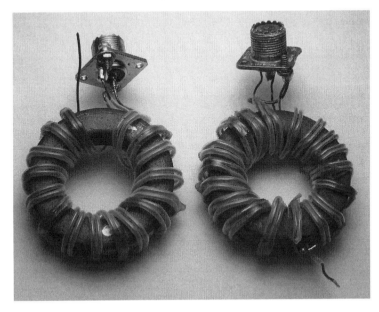

Figure 9-9   *Two 1:4 baluns for the 50:200 Ω level are shown: (Left) Ruthroff design. (Right) Guanella design with both transmission lines on the same core.*

Right: Each of the two transmission lines of this Guanella balun has nine bifilar turns of the same wire and insulation and toroid as in the Ruthroff balun on the left in the figure. Since the total number of turns is now 18, an extra margin of 50% exists in the low frequency response over the Ruthroff balun. The coax connector is also on the low impedance side.

Figure 9-10 shows a comparison in the frequency response between these two baluns for three different values of R_L. These results highly favor the Guanella balun and really don't tell the whole story. At the optimum impedance level of 50:200 Ω (for both baluns), the Ruthroff balun showed appreciable phase shift beyond 1.5 MHz, while the Guanella balun showed virtually no phase shift even up to 100 MHz. And finally, the 4C4 toroid from Ferroxcube was selected since it does not have the failure mechanism with high flux density that practically all of the other ferrites possess. It is the ferrite of choice where possible abuse (as in an antenna tuner) can take place.

Figure 9-11 is a photograph of two other 1:4 baluns that supply further information regarding their use in antenna tuners. These two baluns were also designed to operate best at the 50:200 Ω level with the following parameters:

Left: This Ruthroff balun approximates the 1:4 balun used in some antenna tuners. It has 16 bifilar turns of no. 14 wire on a T-200, 2 in OD, powdered

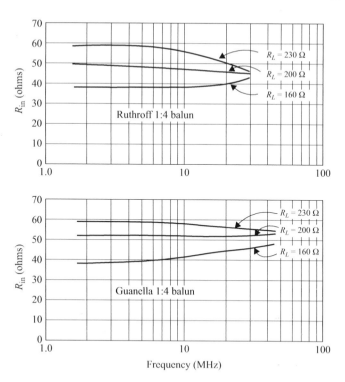

Figure 9-10   These plots show a comparison of the frequency response between the two 1:4 baluns from Figure 9-9.

Figure 9-11 *Photo used to study the relative merits of (left) a transformer using powered iron toiroid, as found in some antenna tuners, and (right) Guanella transformer using a single ferrite core.*

iron toroid ($\mu = 10$). One of the wires is covered with 17 mil wall Teflon tubing. Since the characteristic impedance of this coiled transmission line is 90 Ω, the impedance level for best high frequency response is 45:180 Ω.

Right: This Guanella balun has nine bifilar turns of no. 16 wire in each of the two transmission lines (thus 18 turns when considering the low frequency response) on a 2.4 in OD, no. 61 toroid ($\mu = 125$). The wires are covered with 17 mil wall Teflon tubing. The characteristic impedance is 105 Ω, and the best high frequency response occurs at the 52.5:210 Ω level. This transformer is capable of handling 1 kW continuous power.

Figure 9-12 shows the frequency responses of these two baluns with various floating terminations, which demonstrates not only the superiority of the Guanella balun with a ferrite core but also the danger of using a powdered iron balun in an antenna tuner. It also shows that the powdered iron balun starts falling off, at the low end, around 7 MHz. This means there is insufficient reactance in the windings to prevent a sizable shunting effect. Below 7 MHz, the transformer becomes inductive. This condition allows for flux in the core. Further, this inductance can become part of the tuned L-C network, resulting in very high currents and flux densities. The Guanella balun in the figure demonstrates an exceptional response from 1.7 to 60 MHz. This balun, which uses the popular 2.4 in OD, no. 61 toroid ($\mu = 125$), should be investigated for possible use in antenna tuners.

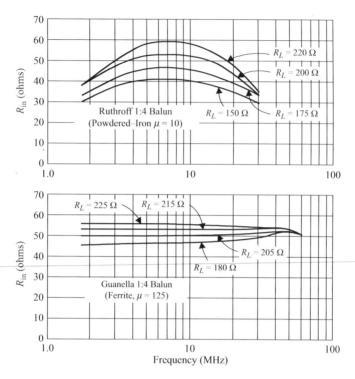

Figure 9-12   These plots show the relative performance of the transformers from Figure 9-11.

9.3.2   75:300 Ω Baluns

The 75:300 Ω balun has been quite popular because of its ability to match 75 Ω coax to the resonant impedance of folded dipoles (which is about 300 Ω). With the use of a very broadband 1:1.56 unun in series, 50 Ω coax can easily be matched to 300 Ω. Figure 9-13 shows two broadband Guanella 1:4 baluns designed to match into balanced, floating loads. The balun on the right is specifically designed to match 75 to 300 Ω. The balun on the left, although optimized for the 50:200 Ω level, is still able to cover three of the amateur radio bands at the 75:300 Ω level because of the short lengths of its transmission lines. The parameters for the two Guanella baluns in Figure 9-13 are as follows:

> *Left*: This Guanella balun, which is optimized at the 50:200 Ω level, has six bifilar turns on the two transmission lines. The wire is no. 16 and is covered with 17 milwall Teflon tubing. The core is a 1 in OD, 250L toroid ($\mu = 250$). The response at the 75:300 Ω level is flat from 1.5 to 10 MHz. Beyond 10 MHz, the impedance ratio increases and becomes complex. This balun performs much better at this level than the larger Guanella balun on

Figure 9-13 *This photo shows two high impedance Guanella transformers that use one core for both transmission lines.*

the right in Figure 9-9 because of its much shorter transmission lines. At the 50:200 Ω level, this balun is essentially flat from 1.5 to 100 MHz. Also, at the 25:100 Ω level this balun still covers the 1.5 to 10 MHz range. The power rating is 1 kW continuous power at the three impedance levels.

Right: This Guanella balun is optimized for the 75:300 Ω level. The spacing between the no. 16 wire (ladder line) is approximately 1/8 inch. The spacers are sections of 3M no. 92 tape. Each transmission line has seven bifilar turns on the 2.4 in OD, no. 64 toroid ($\mu = 250$). The response is flat from 3.5 to 50 MHz. If operation down to 1.5 MHz is required, then a 2.62 in OD, K5 toroid ($\mu = 290$) and nine bifilar turns is recommended. A permeability of 125 would probably raise the efficiency from 97 to 98%. A permeability of 40 would raise it to 99%. The same comment can be made for the Guanella balun on the left in Figure 9-13, but lowering the permeability affects (negatively) the low frequency response proportionally.

9.3.3   25:100 Ω Baluns

Although not nearly as popular as the baluns described previously, the 25:100 Ω balun does have some applications. For example, if it is in series with a 50:25 Ω unun, then a broadband match can be made from a 50 Ω coax to the balanced and floating impedance of a quad antenna. This compound arrangement can also be done on a single toroid and is described in section 9.5.2. A test balun, which is actually optimized at the 27.5:110 Ω level, has an essentially flat 1:4 ratio from 1.5 to 100 MHz with loads varying from 90 to 120 Ω. It has six bifilar turns of no. 16 wire (held closely together) in each of its two transmission lines. The core is a 1 1/2 in OD, no. 64 toroid ($\mu = 250$). The first version of this balun had the same number of bifilar turns, but they were spaced much closer to each other. The best high frequency response occurred at the 25:100 Ω level. This balun showed appreciable phase shift with a resistive bridge (and hence roll-off) at 50 MHz. By simply increasing the spacing between adjacent bifilar turns (particularly on the inside diameter of the toroid), the high frequency response more than doubled. Virtually no phase shift was observed at 100 MHz, which was the limit of the bridge. This was a clear demonstration of the very high frequency capability of the Guanella balun when the parasitic elements are considerably reduced. The Ruthroff balun, because it adds a delayed voltage to a direct voltage, cannot approach this performance when the load is floating.

9.3.4   12.5:50 Ω Baluns

Very broadband baluns matching 50 Ω coax cable to balanced loads of 12.5 Ω (floating or grounded at their midpoints) are readily made using Guanella's approach. In this case, the high side of the transformer is grounded—that is, terminal 2 in the Guanella balun (Figure 9-7b) instead of terminal 1. These transformers can be designed to maintain their 1:4 impedance ratios over a very wide bandwidth with loads that vary from 9 to 15 Ω. This is the range for many Yagi beam antennas. These baluns can also be made to handle more than 1 kW continuous power. By using low impedance coax cable or polyimide coated wire like ML or H Imideze, they can withstand several thousand volts without break-down. Two of these baluns are shown in Figure 9-14. The coax connectors are now on the low side of the baluns. The parameters for these three baluns are as follows:

> *Left*:12 bifilar turns of no. 14 wire, tightly wound on 3/8 in diameter rods, 3 in long, $\mu = 125$ ferrite. The impedance ratio, when matching 50 Ω (unba-lanced) to 12.5 Ω (balanced and floating), is essentially flat from 1.5 to over 30 MHz. With the center of the 12.5 Ω load grounded, the range is 3 to over 30 MHz. The power rating is 1 kW continuous power. If 1/2 in diameter rods were used, then 10 bifilar turns would give the same performance. The length of the rod, which is not especially critical, can vary from 3 to 4 in.
>
> *Right*: 7 1/2 turns of low impedance coax cable on 1/2 in diameter, inch long rods of $\mu = 125$ ferrite. The inner conductor of no. 12 wire has two layers of 3M no. 92 tape. The outer braid, which is unwrapped, is from RG-122/U cable (or equivalent). At the 11.75:47 Ω level (which is the optimum level), the response with a floating load is flat from 3.5 to well over 30 MHz. With the center of the load grounded, it is flat from 7 to well over 30 MHz. These numbers are duplicated at the 12.5:50 Ω level. The power rating is over 2 kW continuous power. The voltage breakdown is in excess of 3000 V if ML or H Imedeze wire is used.

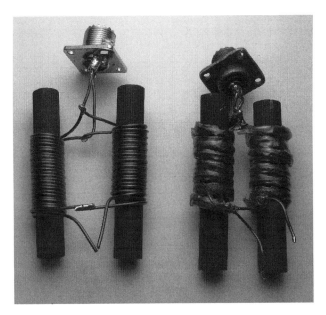

*Figure 9-14   Photo shows two low impedance Guanella transformers:
(Left) Parallel line. (Right) Low impedance coax.*

9.4   The 1:9 Balun

When matching a 50 Ω coax down to a balanced load of 5.56 Ω or up to a balanced load of 450 Ω, the Guanella balun is the transformer of choice. There is little doubt that these baluns offer the widest bandwidths under these two very different conditions. Further, this modular concept (i.e., adding transmission lines in parallel-series arrangements) offers the highest efficiency at high impedance levels. Experiments have shown that efficiency, at least with the Ruthroff unun, decreases as the impedance level increases. With Guanella's approach, each transmission line shares a portion of the load; therefore, his transmission lines can work at lower impedance levels. Also, the longitudinal gradients are less with his transformers.

The balun in Figure 9-15 is designed to match 50 to 5.56 Ω. This transformer could be used to match 50 Ω coax cable directly to short-boom, four-element Yagi beams with resonant impedances of about 6 Ω. This low impedance 1:9 balun has turns of low impedance coax ($Z_0 = 13$ Ω) on each of the three ferrite rods ($\mu = 125$). The rods have a diameter of 1/2 in and a length of 4 in.

Figure 9-16 is a photograph of a 1:9 balun capable of broadband operation from 450 to 600 Ω. Each of the three 2.4 in OD, no. 64 toroids ($\mu = 250$) has 11 bifilar turns of 300 Ω TV ribbon. Due to the proximity of the turns on the inside diameters of the cores, the characteristic impedance is lowered to 205 Ω. At the 68.33:615 Ω, level, which is optimum, the response is flat from 5 to 40 MHz. By using a larger core, one to two more turns would be possible, thereby extending the frequency

*Figure 9-15   Photo shows a 1:9 low impedance transformer designed for the
50:5.56 Ω range.*

*Figure 9-16   Photo illustrates a 66.7:600 Ω transformer constructed using 300 Ω
twin lead.*

range to 3.5 to 40 MHz. At the 66.67:600 Ω impedance level, the high frequency
response of these transformers is still above 30 MHz. The power rating is at least
500 W continuous power. Further, by adding a 1:1.36 unun and a 1:1 balun in series
(and this can be done with one core; see Chapters 7 and 8), this compound arrangement becomes an excellent unun for matching 50 Ω (unbalanced) to 600 Ω

(unbalanced). Sevick found it necessary to use a hole punch to remove material between the wires to allow the twin lead to bend appropriately. Figure 9-16 was created using twin lead from a GQ brand FM dipole antenna, and the cable was flexible enough to easily wind around the cores. Flexibility will depend on the wire gauge of the twin lead and the thickness of the insulation.

The high frequency and low frequency models of the low impedance transformer shown in Figure 9-15 are presented in Figure 9-17a and Figure 9-17b, respectively. The inner conductors are no. 12 wire with two layers of 3M no. 92 tape. The

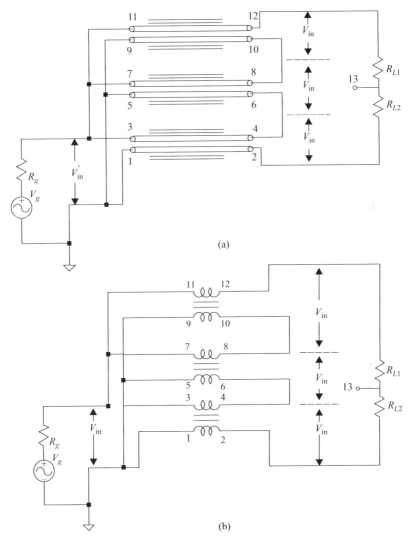

Figure 9-17 *(a) High frequency model of the Guanella 1:9 balun. (b) Low frequency model. It is assumed that $Z_0 = R_L/3$.*

outer braids (unwrapped) are from RG-122/U cable (or equivalent). At the 5.56:50 Ω level, the impedance ratio (with the load floating) is constant from 1.5 MHz to over 30 MHz. The power rating is in excess of 2 kW continuous power. With ML or H Imideze wire, the voltage breakdown is in excess of 3000 V. Although more awkward to construct, four turns of the same coax on toroids with permeabilities of 250 to 300 would yield a 1:9 balun with much greater bandwidth. Finally, a broadband 1:16 balun could be constructed with four coax cables using no. 10 wire with one layer of 3M no. 92 tape for the inner conductor. The characteristic impedance of this coax would be about 9 Ω. This balun would match 3.125 Ω to 50 Ω.

9.5   Baluns for Yagi, Quad and Rhombic Antennas

A very popular balun for antenna use has been the 1:1 (50:50 Ω, nominally) trifilar design by Ruthroff. It has been used successfully in matching 50 Ω coax to Yagi beams after shuntfed methods were employed to raise the input impedance. It has also found success in matching 50 Ω coax directly to 1/2 λ dipoles at heights of 0.15 to 0.2 λ, where the resonant impedances are 50 to 70 Ω, respectively (the resonant impedance reaches a peak of about 98 Ω at a height of 0.34 λ). Outside of these two cases, baluns have found very little use in matching 50 Ω coax cable to resonant impedances far removed from the "nominal" 50 Ω. For the experimenter, baluns for the following antennas are offered.

9.5.1   Yagi Beams

Sections 9.3 and 9.4 described Guanella baluns with ratios of 1:4 and 1:9 that can match 50 Ω coax directly to Yagi beams with balanced and floating impedances of about 9 to 15 Ω and 5 to 8 Ω, respectively. Sevick designed two other baluns capable of matching 50 Ω coax directly to higher impedance Yagi antennas. One balun is designed to match 50 Ω coax to a balanced (and floating) impedance of about 20 Ω. Its useful impedance range is probably from 16 to 25 Ω. The schematic is shown in Figure 9-18. It is a compound transformer consisting of a step-down (50:22.22 Ω) Ruthroff-type unun in series with a low impedance coax cable 1:1 (22:22 Ω) Guanella balun. The core is a 2 in OD, no. 61 toroid ($\mu = 125$), and both transformers are

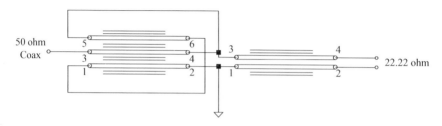

Figure 9-18   Schematic shows a fractional ratio balun designed to match 50 Ω cable to a balanced floating load near 20 Ω. The 1:1 balun on the right uses 22 Ω coax cable.

wound on the same core. The unun has five trifilar turns of no. 14 wire. The 1:1 coax cable balun also has five turns. The coax cable uses no. 12 wire with two layers of 3M no. 92 tape for the inner conductor. The outer braid, which is left untaped, is from RG-122/U. At the 50:22.22 Ω level, the response is flat from 3.5 to well beyond 30 MHz. The power rating is in excess of 1 kW continuous power. If 160 meter operation is desired, then a core with a permeability of 250 to 300 is recommended.

The other balun is designed to match 50 Ω coax to a balanced and floating impedance of about 30 Ω. Its useful impedance range is probably from 25 to 35 Ω. It is also a compound transformer using a step-down (50:28.13 Ω) Ruthroff-type unun in series with a low impedance coax cable 1:1 (30:30 Ω) Guanella balun (Figure 9-19). The common core is a 2.4 in OD, no. 64 toroid ($\mu = 250$). The unun, which has an impedance ratio of 1.78:1, has six quadrifilar turns. Winding 5–6 in Figure 9-19 is no. 14 wire, and the other three are no. 16 wire. The 1:1 coax cable balun also has six turns. The inner conductor of the coax cable is no. 14 wire with two layers of 3M no. 92 tape, followed with two layers of 3M no. 27 glass tape. The outer braid (untaped) is from RG-122/U cable. At the 50:28.13 Ω level, the response is flat from 1.5 to 50 MHz. The power rating is in excess of 1 kW of continuous power. This transformer, as well as the previous one, could also have been constructed with two separate cores.

9.5.2 Quad Antennas

The quad antenna generally has a balanced (and floating) resonant impedance in the range of 100 to 120 Ω. This antenna also lends itself readily to a compound balun. Several approaches can be used; for example, a 1:2 step-up unun (50:100 Ω) followed by a 1:1 balun (100:100 Ω) or a 2:1 step-down unun (50:25 Ω) followed by a 1:4 step-up balun (25:100 Ω). Sevick tried the latter approach, and it was implemented as a compound balun using a single 2.4 in OD, no. 61 toroid ($\mu = 125$) (Figure 9-20). It uses a tapped trifilar step-down unun in series with a Ruthroff 1:4 balun. The Guanella 1:4 balun, although possessing a better high frequency response, was not used since it did not lend itself as readily to a single core. If a much wider bandwidth is required, then two separate cores, with the 1:4 balun using Guanella's approach, is recommended. The unun in Figure 9-20 has six trifilar turns.

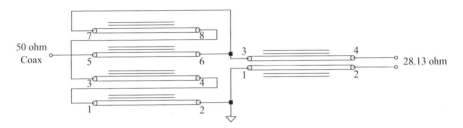

Figure 9-19 Schematic shows a fractional ratio balun designed to match 50 Ω cable to a balanced floating load near 30 Ω. The 1:1 balun on the right uses 30 Ω coax cable.

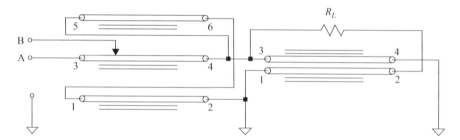

*Figure 9-20   Schematic shows the connections for a compound 1:2 step-up balun.
With the input at terminal B, the impedance ratio is 1:2 (50:100 Ω).
With the input at terminal A, the impedance ratio is 1:1.78 (50:90 Ω).*

Winding 3–4 is no. 14 wire and is tapped at one turn from terminal 3. The other two windings are no. 16 wire. With the input connection to the tap, the impedance ratio is 2:1. The Ruthroff 1:4 balun uses 10 bifilar turns of no. 14 wire. This compound balun, 50 Ω coax to 100 Ω (balanced), is flat from 3.5 to 30 MHz. The response is quite the same in matching 60 Ω (unbalanced) to 120 Ω (balanced). With the input connection directly to terminal 3, a similar response is obtained at the 50:90 Ω impedance level. If 160 m operation is also desired, then a toroid with a permeability of 250 to 290 is recommended. The power rating is 1 kW continuous power.

9.5.3   Rhombic Antennas

Compound baluns also lend themselves to matching 50 Ω coax to the balanced (and floating) resonant impedances of V and rhombic antennas. These impedances are generally in the range of 500 to 700 Ω. Section 9.4 described some compound baluns using ununs in series with Guanella baluns, yielding wideband responses over this impedance range. This subsection presents one of Sevick's earlier approaches for a 1:12 balun using only two toroidal transformers. One is a tapped bifilar Ruthroff unun with a ratio of 1:3, and the other is a Ruthroff 1:4 balun. Figure 9-21 shows the performance when matching to a balanced (and floating) load of 600 Ω. In this case, the input impedance is measured as a function of frequency. As shown, a constant ratio is obtained in the frequency range of 7 to 30 MHz. Figure 9-22 shows the schematic for the two-series transformers. The transformer on the left has eight bifilar turns of no. 14 wire on a 2.4 in OD, Q1 toroid ($\mu = 125$). The wire is covered with 17 mil wall Teflon tubing. The top winding in Figure 9-22 is tapped six turns from terminal 3, giving a 1:3 step-up ratio. The transformer on the right has 11 bifilar turns of no. 16 wire on a 2.4 in OD, Q2 toroid ($\mu = 40$). The wire is also covered with 17 mil wall Teflon tubing. This is a 1:4 Ruthroff balun. Although Q2 material has a lower permeability than Q1 material, it was chosen because it has a lower core loss at these high impedance levels.

This combination of transformers allows for broadband operation at the high impedance levels of 500 to 700 Ω because of the canceling effects they have in a series configuration. Since the characteristic impedance (Z_0) of the 1:4 balun on the right is only about 130 Ω (it should be 300 Ω for optimum response), the input

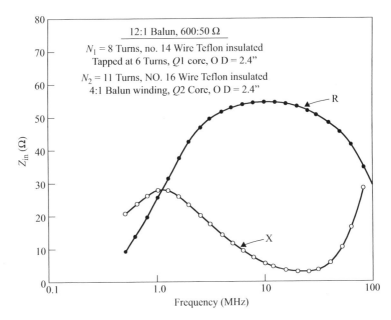

Figure 9-21   *Plots show the performance of a 1:12 balun designed to match 50 Ω coax cable to a balanced load of 600 Ω.*

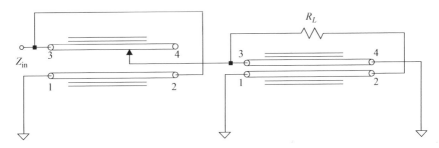

Figure 9-22   *Schematic shows the connections of the 1:12 combination balun measured in Figure 9-21.*

impedance as seen at its terminals 1–3 is capacitive and the real part is less than $R_L/4$. Since the characteristic impedance (Z_0) of the tapped transformer on the left is about 115 Ω and is greater than would be normally used to match 50 to 150 Ω, it has the opposite effect on its load. It causes the load to look inductive; the reactance of the right-hand transformer is effectively canceled over a large portion of the band. The resistive component is not altered when the characteristic impedance is greater than would normally be used.

Even though the compound baluns described in section 9.4 (using Guanella's approach) have the potential for much wider bandwidths, the two octaves achievable using the simple schematic in Figure 9-22 should prove to be quite useful.

Chapter 10

Multimatch Transformers

10.1   Introduction

Multiband vertical antennas have enjoyed considerable popularity with radio amateurs (including Sevick) because of their low angle of radiation and flexibility in changing bands without any form of external switching. Most of these multiband verticals have high degrees of inductive loading and thus very narrow bandwidths on the 80 and 160 m bands.

To obtain greater bandwidths on these bands and still maintain a multiband antenna system with a high degree of vertical radiation, Sevick connected a four-band (10–40 m) trap vertical in parallel with sloper and inverted-L antennas for the 80 and 160 m bands. This was done over a low loss ground system of 100 radials, each about 50 to 60 ft long. Impedance measurements showed that the sloper and inverted-L antennas had resonant impedances of 12 to 25 Ω, depending on (1) the slope angle of the sloper, (2) the height of the vertical portion of the inverted-L, and (3) the interaction between the 80 and 160 m antennas. Since the inverted-L antennas were mounted 8 to 12 in from the trap vertical, little difference was noted in the input impedance of the trap vertical or its performance. Slopers had very little effect since the capacitive coupling was minimal.

Instead of using separate feed lines and matching transformers or a single feed line and relays that switch to the appropriate matching transformers, parallel transformers were investigated for possible use. Two step-down transformers with different ratios were connected in parallel on their 50 Ω sides and to various antennas on their output sides. Since the mutual coupling of parallel connected transmission line transformers was found to be minimal (like putting a short length of unterminated transmission line across their inputs), this arrangement pretty much duplicated the well-known technique of feeding parallel connected dipoles for different bands with a single coax. This process worked well for two different impedances but left something to be desired if three or four broadband ratios were required.

This led to an investigation for obtaining two broadband ratios from a single transformer. Transformers capable of supplying either 1:1.5 and 1:3 or 1:2 and 1:4 ratios were then successfully designed. Again, by connecting these two transformers at their 50 Ω sides, which were the high impedance sides, four broadband ratios now became available. Although this technique was used only for matching

50 Ω coax to lower impedances, it should also work in matching 50 Ω coax to a variety of both high and low impedances.

10.2   Dual-Output Transformers

Chapters 7 and 8 introduced the concept of higher order windings and how broadband ratios of less than 1:4 could be obtained with trifilar, quadrifilar, and higher order transformers. Of particular significance was the transposition of the various windings to obtain optimum characteristic impedances for either rod or toroidal transformers. Most of these transformers could yield more than one impedance ratio by either tapping a winding or by direct connections to the terminals of the inner windings. Generally these transformers were optimized for a single impedance ratio. This section introduces the concept of transposing the windings such that two broadband ratios become available. As will be seen, the schematics are considerably different from those of the two earlier chapters. In general, these transformers do not quite exhibit the high frequency response for both ratios of a single ratio transformer. This is because it is difficult to optimize equally the characteristic impedances of the windings for two different broadband ratios. But in most cases, the two impedance ratios can be found to be constant from 1.5 to 30 MHz. In many cases, the high frequency response of one of the ratios easily exceeds 30 MHz.

10.2.1   1:1.5 and 1.3 Ratios

Although the actual ratios obtained by the following transformers are 1:1.56 and 1:2.78, respectively, very little difference will be noted from ratios of 1:1.5 and 1:3 when matching to antennas. This is because of the variation of their input impedances with frequency. A slight shift in the best match point (lowest VSWR) with frequency might be observed. Figure 10-1a shows the basic schematic of a dual-output, quintufilar, unun transformer. Terminal C is the high impedance side. Figure 10-1b and Figure 10-1c, which are toroidal and rod versions, are specifically designed to match 50 Ω coax to lower impedances. The two broadband impedance ratios, which are the same for all three transformers, can best be determined from the basic schematic of Figure 10-1a.

If an input voltage (V_{in}) is connected from terminal A to ground, the four transmission lines have $V_{in}/4$ on their inputs. Terminals 2, 4, 6, and 8 are all bootstrapped with a potential of $V_{in}/4$ by their connections to the odd-numbered terminals on the left. The voltage between terminals 10 and 8 came from $V_{in}/4$, on the left, which traversed a transmission line. Thus, the voltage at terminal C becomes $5/4V_{in}$, and the impedance ratio becomes $(V_{out}/V_{in})^2 = (5/4)^2 = 1.56$. The top winding, 9–10, carries four-fifths of the current into terminal A and the other four only one-fifth of the current.

If an input voltage (V_{in}) is connected from terminal B to ground, then the three bottom transmission lines have $V_{in}/3$ on their inputs. Terminals 2, 4, and 6 are now bootstrapped by a potential of $V_{in}/3$ by their connections to the odd-numbered terminals on the left side. The voltage between terminals 8 and 6 came from $V_{in}/3$,

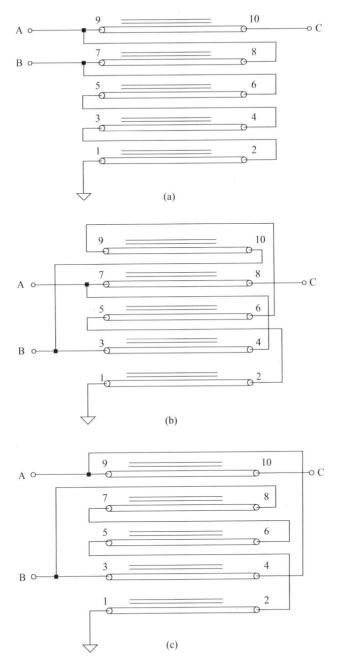

Figure 10-1   Schematic shows three versions of a combination quintufilar 1:1.156 and 1:2.78 transformer: (a) Basic configuration. (b) Version optimized for toroidal core use. (c) Version optimized for rod use.

on the left, which traversed a transmission line. The voltage between terminals 10 and 8 came from the same voltage, which traversed the transmission line twice. Thus, the output at C becomes 5/3 V_{in}. The impedance ratio then becomes $(V_{out}/V_{in})^2 = (5/3)^2 = 2.78$. The two top windings in Figure 10-1A carry three-fifths of the current into terminal A and the other three only two-fifths.

Sevick created three low impedance toroidal versions (Figure 10-1b) of the quintufilar, dual-output transformer. The coax chassis connectors are all on the high impedance sides of the transformers. The parameters of these three transformers are as follows:

(A)   Five quintufilar turns on a 1 1/2 inch OD, no. 64 toroid ($\mu = 250$). Windings 7–8 and 3–4 are no. 14 wire, and the other three are no. 16 wire. When the input is connected to terminal A (the 1:1.56 ratio), the response is flat from 1.5 to 30 MHz at the 32:50 Ω level. At the optimum impedance level of 29:45 Ω, the response is flat from 1.5 to 45 MHz. When the input is connected to terminal B (the 1:2.78 ratio), the response is flat from 1.5 to 45 MHz at the 18:50 Ω level. At the optimum impedance level of 20:56 Ω, the response is flat from 1.5 to over 50 MHz. The power rating with either ratio is 1 kW continuous power.

(B)   Five quintufilar turns on a 1 1/4 in OD, K5 toroid ($\mu = 290$). Windings 7–8 and 3–4 are no. 16 wire and the other three are no. 18 wire. When the input is connected to terminal B (the 1:2.78 ratio), the response is flat from 1 to 45 MHz at the 18:50 Ω level. This is at the optimum impedance level. When matching at the 32:50 Ω level, using terminal A, the response is flat from 1 to well over 50 MHz. This is also at the optimum impedance level. This transformer would easily handle the power from any popular transceiver. Its power rating for either ratio is in excess of 200 W continuous power.

(C)   This dual-output transformer was designed for 75 Ω operation (on the high side). It has 4 quintufilar turns on a 1 1/2 in OD, no. 64 toroid ($\mu = 250$). Windings 7–8 and 3–4 are no. 14 wire, and the other three are no. 16 wire. Winding 7–8 is covered with 17 mil wall Teflon tubing. When using terminal A (the 1:1.56 ratio) and matching 50 Ω to 75 Ω, the response is flat from 1.5 to 30 MHz. It is also the same at the 32:50 Ω level. The optimum response occurs at the 38.5:60 Ω level. Here, it is flat from 1.5 to over 50 MHz. When using terminal B (the 1:2.78 ratio) and matching 27 Ω to 75 Ω, the response is flat from 1.5 to well over 50 MHz. This ratio of 1:2.78 also works well at the 18:50 Ω level. At this level the ratio is constant from 1.5 to 30 MHz. The reason for the good performance over these wide impedance levels is because of the short lengths of the transmission lines—only 7 1/2 in long. The power rating with either ratio is 1 kW continuous power.

Sevick also created two low impedance rod versions (Figure 10-1c) of dual-output ununs. A third input connection is also made to terminal 7, yielding a third ratio (but with very much less bandwidth) of 1:6.25. Both of these transformers are rated at 1 kW continuous power with all ratios. The parameters for these two transformers are as follows:

(D) Seven quintufilar turns on a 1/2 in diameter, 3 3/4 in long ferrite rod ($\mu = 125$). Windings 9–10 and 3–4 (Figure 10-1c) are no. 14 wire. The other three are no. 16 wire. Winding 9–10 is covered with 17 mil wall Teflon tubing. When using terminal A (the 1:1.56 ratio) and matching 32 Ω to 50 Ω, the response is flat from 1.5 to about 30 MHz. At 30 MHz, the ratio becomes a little less than 1:1.56. This 32:50 Ω level is the optimum impedance level. When matching at the 18:50 Ω level, using Terminal B in Figure 10-1c, the response is flat from 1.5 to 30 MHz. At the optimum impedance level of 21.5:60 Ω, it is flat from 1.5 to over 45 MHz. When matching with the 1:6.25 ratio (connecting the input to terminal 7) at the 8:50 Ω level, the response is flat from 1.5 to 10 MHz.

(E) Nine quintufilar turns on a 3/8 in diameter, 3 3/4 in long ferrite rod ($\mu = 125$). Windings 9–10 and 3–4 (Figure 10-1c) are no. 14 wire. The other three are no. 16 wire. Winding 9–10 is covered with four layers of 3M no. 27 glass tape (14 mil of insulation). This transformer also has a third input at terminal 7. The response and power rating pretty much duplicate that of the 1/2 in diameter rod transformer in (D).

Two dual-output, broadband transformers for mounting at the base of verticals or slopers and inverted-L antennas were tested. The rod transformer is identical in construction, performance, and power rating to the seven-turn quintufilar rod transformer in (D). The only difference is that winding 9–10 (Figure 10-1c) is covered with two layers of 3M no. 27 glass tape (14 mil of insulation) instead of 17 mil wall Teflon tubing. The performance of this second toroidal transformer is identical to the 5-turn quintufilar toroidal transformer in (A). The power rating is also the same. The only difference in construction is that the core is a 1 1/2 in OD, 4C4 toroid ($\mu = 125$). Since its cross sectional area is about twice that of the no. 64 toroid ($\mu = 250$) used in (A), the low frequency responses are practically identical. The same is true with the high frequency responses, since the transmission lines of the two transformers are not significantly different in length.

10.2.2   1:2 and 1:4 Ratios

In working with ground-fed, multiband antennas systems (over a good ground system) employing combinations of verticals, slopers, and inverted-L antennas, a need arose for low impedance ratios other than 1:1.5 and 1:3. Some antennas had resonant impedances near 12 Ω and others near 25 Ω. Therefore, a study was undertaken to achieve two broadband ratios of 1:2 and 1:4 with a single core. The quadrifilar transformer was found to yield two broadband ratios of 1:1.78 and 1:4. The 1:1.78 ratio generally satisfies the 1:2 ratio requirement. Figure 10-2 shows the schematic of the basic quadrifilar winding (used for analysis purposes) and the final

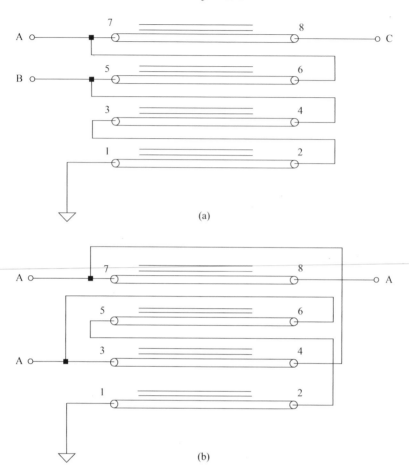

Figure 10-2   Schematic shows the connections of quadrifilar windings for
obtaining broadband ratios of 1:1.78 and 1:4. (a) Used for analysis
purposes. (b) Used with both rods and toroids in matching 50 Ω coax
at terminal C to 28 Ω at terminal A or 12.5 Ω at terminal B.

design (Figure 10-2b) for both rod and toroidal versions. The impedance ratios are
determined from Figure 10-2a as follows:

Top: If an input voltage (V_{in}) is connected from terminal B to ground, the bottom
two transmission lines have $V_{in}/2$ on their inputs. Terminals 2 and 4 are
bootstrapped by a potential of $V_{in}/2$ by their connections to terminals 3 and 5,
respectively. Terminals 6 and 8 each have an added voltage of $V_{in}/2$ as a result
of a voltage of $V_{in}/2$ traversing transmission lines. The added $V_{in}/2$ at terminal
8 traverses the transmission line twice since terminal 7 is connected to terminal 6. Thus, the voltage at terminal C then becomes $2V_{in}$, resulting in an
impedance ratio of 1:4. Since the $V_{in}/2$ voltage from the top transmission line
has twice the delay of the middle transmission line, the high frequency

response with the 1:4 ratio is not as good as the 1:1.78 ratio. By using small, high permeability toroids (250 to 300), it is possible to cover 1.5 to 30 MHz with the 1:4 ratio and still handle 1 kW continuous power.

Bottom: If an input voltage (V_{in}) is connected from terminal A to ground, the three transmission lines have $V_{in}/3$ on their inputs. Terminals 2, 4, and 6 are then bootstrapped by a potential of $V_{in}/3$ by their connections to the odd-numbered terminals on the left. The voltage between terminals 8 and 6 is from the input $V_{in}/3$, which traverses the top transmission line, and thus the voltage at C becomes $4/3V_{in}$. Hence, the impedance ratio becomes $(V_{out}/V_{in})^2 = (4/3)^2 = 1.78$. The top winding 7–8 carries three-fourths of the current into terminal A and the other three only one-fourth of the current.

Sevick created two versions of dual output transformers with ratios of 1:1.78 and 1:4. Both of these transformers are capable of handling 1 kW continuous power with either of their ratios. The chassis connectors are on the 50 Ω side. The parameters for the two transformers are as follows:

(a)   Eight quadrifilar turns on a 1/2 in diameter, 3 3/4 in long rod ($\mu = 125$). Windings 7–8 and 3–4 are no. 14 wire. The other two are no. 16 wire. At the 1:4 ratio (input at terminal B), the response is essentially flat from 1.5 to 21 MHz when matching 12.5 Ω to 50 Ω. At 30 MHz, the ratio increases some and also becomes complex. The optimum response (which is not much better) occurs at the 15:60 Ω level. At the 1:1.78 ratio (input at terminal A), the response is flat from 1.5 to 30 MHz when matching 28 Ω to 50 Ω. This is also the optimum impedance level. This transformer again demonstrates the usefulness of rod transformers in amateur radio.

(b)   Four quadrifilar turns of no. 14 wire on a 1 1/2 in OD, K5 toroid ($\mu = 290$). The turns are crowded to one side of the toroid to lower the characteristic impedance of the windings. At the 1:4 ratio (input at terminal B), the response is essentially flat from 1.5 to 30 MHz when matching 12.5 Ω to 50 Ω. The optimum response occurs at the 15:60 Ω level, where it is flat from 1.5 to 45 MHz. At the 1:1.78 ratio (input at terminal A), the response is flat from 1.5 to 45 MHz when matching 28 Ω to 50 Ω. At the optimum impedance level of 34:60 Ω, the high frequency response is well over 60 MHz. This is probably the best dual-output transformer Sevick constructed and has ratios of 1:1.78 and 1:4.

10.3   Parallel Transformers

As noted in the introduction to this chapter, two transmission line transformers can be connected in parallel on their input sides (usually the 50 Ω sides) and still offer their broadband ratios. The transformer that sees its proper match takes the load while the other one is essentially transparent. Sevick has constructed step-down, parallel-connected transformers matching a single 50 Ω coax to the various lower impedances of vertical, sloper, and inverted-L antennas. This technique is akin to feeding parallel-connected dipoles (on separate bands) with a single coax. The small interaction between the transformers is due to the parasitic capacitance (that of a short

Table 10-1   Input Impedance and Ground Loss of a Resonant
            1/4 λ Vertical Antenna with Number of 1/4 λ Radials

No. of Radials	Input Impedance (Ω)	Ground Loss (Ω)
1 (or ground rod)	85	50
4	65	30
8	57	22
40	39	4
100	36	1

transmission line) of the floating transformer. Since toroidal transformers require fewer turns and hence have shorter transmission lines, their parasitic capacitances are lower than those of rod transformers. But at the 50 Ω input level, rod transformers have also been found to be acceptable in the frequency range of 1.5 to 30 MHz.

This technique of parallel-connected transformers should also find use with other antenna systems. For example, a step-up transformer and a step-down transformer (from 50 Ω) should also work as well. By using 1:1 baluns on the outputs, beams and dipoles can also be fed from a single coax. Further, two antennas designed for the same frequency, but with different impedances, can also be matched. In this case, the input impedance of the parallel-connected transformers becomes 25 Ω. A series 2:1 unun can then bring the impedance back to 50 Ω.

It was also noted in the introduction that parallel-connected transformers have been used with various ground-fed antennas over a low loss radial system (100 radials of 50–60 ft in length). The resonant antenna impedances seen by the transformers were between 12 and 35 Ω. If fewer radials are used, the added loss due to a poorer ground system has to be taken into account. As a reminder, Table 10-1 provides information on loss versus the number of 1/4 λ ground radials as well as the resonant input impedance of a 1/4 λ vertical with the added loss. It is safe to assume that the loss figures should also apply to ground-fed sloper and inverted-L antennas. As can be seen from Table 10-1, the multimatch transformers described in this chapter should be mainly used with ground systems of 40 or more 1/4 λ (or longer) radials. On 160 m, the 50–60 ft radials (100 of them) added only a few ohms of extra loss.

An example of transformers operating in parallel is shown in Figure 10-3, which is a schematic of two broadband, single-ratio transformers, connected in parallel on their 50 Ω (high) sides. The parameters for these two transformers, which are capable of handling 1 kW continuous power, are as follows:

Top: This 1:1.56 ratio transformer has five quintufilar turns on a 1 3/4 in OD, K5 toroid ($\mu = 290$). The center winding 5–6 is no. 14 wire. The other four are no.16 wire. At the 32:50 Ω level, which is optimum, the response is flat from 1.5 to well beyond 30 MHz.

Bottom: This 1:2.25 ratio transformer has eight trifilar turns on a 1 1/2 in OD, 4C4 toroid ($\mu = 125$). The center winding 3–4 is no. 14 wire. The other two are no. 16 wire. At the 22.22:50 Ω level, which is optimum, the response is flat from 1.5 to well beyond 30 MHz.

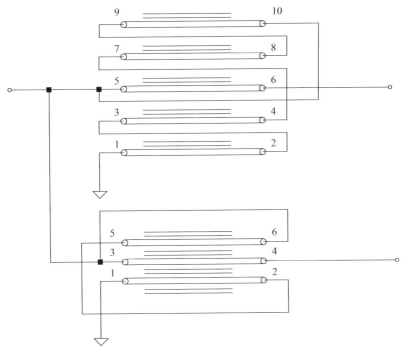

Figure 10-3   *Schematic shows two transformers connected in parallel at the 50 Ω*
input and separate connections to different loads: (Top) 1:1.156.
(Bottom) 1:2.25.

10.4   Eight-Ratio Transformer

Sevick's early work produced an unun transformer with eight separate ratios using
a single toroidal core. Although an improved version is possible (and will be discussed later in this section), the earlier version is reproduced here because it uses a
popular toroid that is probably available in many radio amateurs' junk boxes.
Figure 10-4 is the schematic, and Table 10-2 lists the performance at various
impedance ratios using 50 and 100 Ω load resistors. This transformer has
6 quadrifilar turns of no. 14 wire on a 2.4 in OD, Q1 toroid ($\mu = 125$). As shown in
Figure 10-4, taps were at two turns from terminal 5 (F) and at five turns from
terminal 5 (E). The useful frequency range in Table 10-2 is defined as the range
where the loss is less than 0.4 dB. This loss at the high frequency end is due to the
transformation ratio becoming complex and increasing or decreasing, depending on
the relationship between the load and the effective characteristic impedance of the
windings. Thus, the 0.4 dB limit at the high end approximates a VSWR of 2:1. This
loss at the high end is not an ohmic loss as such but is because of the load's inability
to absorb the full available power. Therefore, the useful ranges listed in Table 10-2
are very different from the ranges quoted on practically all of the other transformers

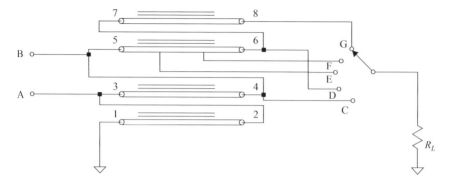

Figure 10-4   Schematic shows an 8 ratio quadrifilar transformer.

Table 10-2   Performance of Quadrifilar Transformer

Input Port	Output Port	Impedance Ratio	Useful Frequency Range	
			RL = 50 Ω	RL = 100 Ω
B	F	1:1.36	1.5 to 30 MHz	3 to 30 MHz
B	E	1:2.0	1.5 to 30 MHz	3 to 30 MHz
B	D	1:2.25	1.5 to 30 MHz	3 to 30 MHz
A	C	1:4.0	1 to 30 MHz	3 to 30 MHz
A	F	1:5.4	1 to 15 MHz	1.5 to 30 MHz
A	E	1:8.0	1 to 15 MHz	1 to 15 MHz
A	D	1:9.0	1 to 15 MHz	1 to 30 MHz
A	G	1:16	1 to 8 MHz	1 to 15 MHz

in this book. For these, the expressions "a flat response" or a "constant impedance ratio" were used between a lower and upper frequency limit, which amounts to a VSWR of 1:1 across almost the entire frequency range.

Several improvements can be made in the eight-ratio transformer. A smaller and higher permeability toroid is recommended—toroids with outside diameters of 1 1/2 to 1 3/4 in and permeabilities of 250 to 300. Because of these changes, fewer turns are needed for adequate low frequency responses. Thus, 4 or 5 quadrifilar turns would be sufficient. Fewer turns (and hence shorter transmission lines) increase the high frequency performance. With fewer turns, the taps would also have to be changed. For the four quadrifilar turns version, the taps should be set at one and three turns from terminal 5 (F and E), respectively. For the five quadrifilar turns version, the taps should be set at one and four turns from terminal 5 (F and E), respectively. The resulting ratios would not differ greatly from those in Table 10-2. And finally, if this transformer is to be used for any length of time in matching 50 Ω to 3.125 Ω (16:1) at the 1 kW level, then winding 1–2 should be replaced with no. 12 or even no. 10 wire. At this power level, the current in winding 1–2 becomes very high since it handles three times more current than the 50 Ω coax. Further, this current tends to crowd between windings 1–2 and 3–4, where the electric field is a maximum.

Chapter 11
Equal Delay Transformers

11.1 Introduction

There are two distinct paths of development for equal delay transformers, each of which can be characterized as achieving a more generalized form of transformer derived from Guanella's earlier research. Although others have explored parallel development, this chapter describes the work of W. A. Lewis at Collins Radio, which was further developed by Blocksome at Collins, and independent work by McClure at RCA [1–5].

11.2 The Need for Fractional Ratio Transformers

McClure's work was specifically directed toward the development of fractional ratio transformers, whereas the work at Collins Radio included both integer and fractional ratio transformers. Although these approaches are different, one of the primary reasons for these efforts was to obtain a wider range of transformation ratios than, for example, the basic 1:4, 1:9, and 1:16, which characterized earlier investigations in this area.

Most transformers explored in the previous chapters are designed for transformation ratios of n^2 or $1/n^2$, where n is an integer. There are, however, many applications where other impedance transformation ratios are desired. Examples include a 1.5:1 ratio for matching between 50 Ω and 75 Ω transmission line systems and 2:1 or 1:2 ratio for matching from 50 Ω to 25 or 100 Ω. Antenna systems and solid-state power amplifiers represent application areas that may require matching a wide range of impedances to a 50 Ω feed system over a bandwidth that can be accommodated only with a properly designed and constructed transmission line transformer.

Chapter 7 described the Ruthroff bootstrap method, which uses multiple windings to provide transformation ratios defined as $(n/m)^2$, where n and m are integers. The quintufilar transformer in Chapter 7 (Figure 7-8) has a transformation ratio of 1:1.56, or $(5/4)^2$, while the trifilar design (Figure 7-20) provides a ratio of 1:2.25, or $(3/2)^2$. Chapters 7, 8, and 10 also explain transformers with tapped windings to obtain fractional transformation ratios. They place a tap at the proper point along the voltage gradient of one winding, which is then summed with the full windings to achieve the desired transformation.

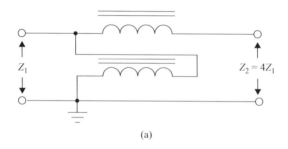

(a)

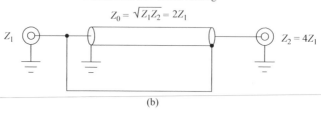

(b)

Figure 11-1   Diagrams show the basic Ruthroff transformer: (a) Low frequency model. (b) Coax line implementation.

All of these fractional ratio transformers require connections that connect one or more delayed voltages from the output side of the transformer to windings at the input side. This is the technique used in the basic Ruthroff 1:4 unun described in Chapter 1. Figure 11-1 shows this transformer in two forms. We will use this circuit as the starting point for the introduction of another class of transmission line transformer, the family of *equal delay* transformers.

By convention, all the transformers illustrated in this chapter are shown with the low impedance connection on the left and the high impedance connection on the right. All ratios are given as numbers greater than 1, such as 1:2.25 or 1:4. However, transformer inputs and outputs are interchangeable, so the same transformers can also provide a step-down in impedance that is the inverse of the ratios shown (e.g., 1:0.44 and 1:2.25, or 1:0.25 and 1:4).

11.3   The Equal Delay Transformer

The Ruthroff transformer in Figure 11-1 requires a connection from the high impedance end of the transformer to the low impedance end. If the transformer is wound on a toroidal ferrite core, the two ends can be brought out very close to one another and the connection made with minimal added length (and stray inductance). In this case, the high frequency limit is established by the length of the transmission line. If it is much less than 1/4 wavelength at the highest frequency of operation, the voltage connected from output to input will have a delay small enough to avoid loss due to signal cancellation caused by excess delay (phase shift).

If we replace the input-to-output connection with another transmission line equal in length to the first, we obtain the transformer in Figure 11-2. In this configuration, the Ruthroff design is modified to create a delay in the connecting line that is equal to the delay in the "main" line, which is where this new configuration gets its name. This is the approach Lewis took.

No ferrite loading is required for the added delay line. There is no voltage drop on the outer conductor; thus, no isolation is required from one end to the other. These lines could be parallel or twisted wires instead of coax cables, but as will be seen this type of transformer is easily described and analyzed using coax transmission lines.

Figure 11-2 also suggests a mechanical advantage. With no connection between the input and output ends of the transformer, there is no need to place the two ends in proximity. Using coax cable, the ferrite loading can readily be implemented with cylindrical cores or stacked toroids, and the opposite ends of the transformer can be separated by the length of the transmission lines, as illustrated in Figure 11-3. This configuration will be practical for many installations where separation of input and output are useful either for maximum isolation or for convenience of layout.

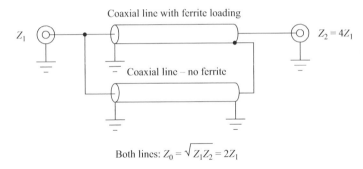

Coaxial line with ferrite loading

Z_1 $Z_2 = 4Z_1$

Coaxial line – no ferrite

Both lines: $Z_0 = \sqrt{Z_1 Z_2} = 2Z_1$

Figure 11-2   A pictorial that shows the connections of a basic equal delay transformer. Replacing the input-to-output connection from the Ruthroff design with another transmission line creates a path with a delay equal to the main line.

Figure 11-3   Photo shows construction of a 4:1 equal delay transformer using a linear configuration and "one turn" of coax cable through many ferrite beads.

11.4   Integer [n^2] Ratio Equal Delay Transformers

The same configuration used for the 1:4 transformer can be extended to include 1:9, 1:16, and higher n^2 impedance ratios. Figure 11-4 shows the connections required for transformers with 1:9 and 1:16 impedance ratios. The top transmission line in the diagram of each transformer experiences the largest voltage drop and thus requires sufficient inductance to achieve adequate isolation, using ferrite sleeves or windings through a toroid core. In the 1:9 transformer, the second line has half the voltage drop of the top wire and requires half the isolation and half the amount of ferrite material or 0.707 the number of turns through a toroid core. As noted with the 1:4 transformer, the shield of the bottom line has no voltage drop and requires no isolation.

The advantage of using coax cable for this transformer is evident. At one end, all shields are grounded, enabling common assembly methods to be used. There is no need to float the shields above ground at this end. Also, the bottom line has its shield grounded at both ends, requiring only the remaining lines at the high impedance end of the transformer to have their shields insulated from ground.

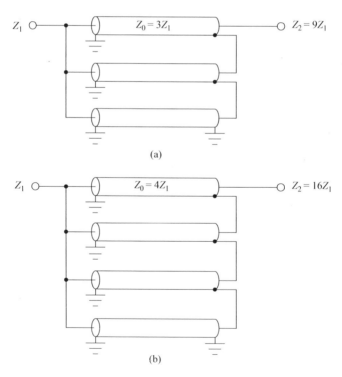

(a)

(b)

Figure 11-4   Connections are shown for implementing equal delay transformers with (a) 1:9 impedance ratio and (b) 1:16 ratio.

11.5 Fractional [$(n/m)^2$] Ratio Equal Delay Transformers

The previous section described transformers where all primary connections are in series and all secondary connections are in parallel. It is also acceptable to combine series and parallel connections on one end, as long as the connections at the other end of the transformer have the opposite connection scheme. Figure 11-5 shows a three-line transformer with a 1:2.25 [$(3/2)^2$] impedance transformation ratio. On the low impedance side, the top line is connected in parallel with the series combination of the lower two lines. At the high impedance side, the top winding is connected in series with parallel connected lower two lines, adhering to the reverse connection requirement.

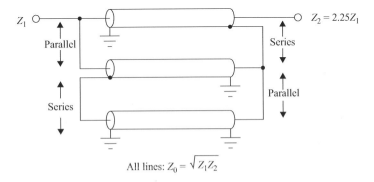

Figure 11-5 *A three transmission line equal delay transformer is shown using series and parallel connections to implement a 1:2.25 impedance ratio.*

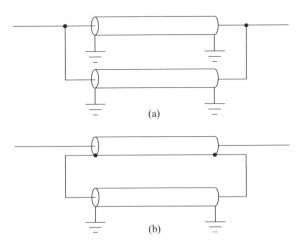

Figure 11-6 *Violations of the "opposite connection" rule create systems with improper operation: (a) 1/2 Z_0 transmission line. (b) Open circuit.*

It is easy to see why the opposite connections are necessary. If the same transmission lines are connected in parallel at both ends, they are no longer two lines: the two lines effectively become a single line with one-half the characteristic impedance. If two lines are connected in series at both ends, continuity is broken and there is no longer an electrical path to the lower lines. Figure 11-6 clearly illustrates this point.

11.6   Unun versus Balun Equal Delay Transformers

The previous examples are all unun equal delay transformers. As such, the top transmission line in each diagram requires the greatest isolation from input to output, with progressively less ferrite loading for the lower lines, ending with no isolation on the bottom line, which has its shield at ground potential at both ends. This structure, however, can be used only as an unun, since isolation is not maintained from input to output on the return path (ground).

If the same arrangement of transmission lines is used, but with adequate ferrite loading to isolate the input and output of each line, we have an equal delay balun transformer. This is the approach McClure takes in his independent development of equal delay transformers.

The equal delay balun is an extension of Guanella's technique. His work was limited to transformers of n^2 impedance transformation ratios. These transformers have all transmission lines connected in parallel at the low impedance side and connected in series at the high impedance side. McClure recognized that Guanella's technique was equally valid for combinations of series and parallel connections, which could produce the much wider range of impedance transformations represented by $[(n/m)^2]$ ratios.

To illustrate the equal delay balun, we can redraw Figure 11-5 as a 1:2.25 impedance ratio balun, shown in Figure 11-7. Now, the input and output of the transformer are isolated with ferrite inductive loading on all lines, while the basic

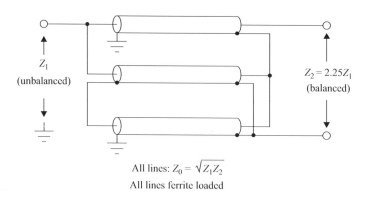

All lines: $Z_0 = \sqrt{Z_1 Z_2}$
All lines ferrite loaded

Figure 11-7   By loading all three lines of Figure 11-5 with ferrites, a balun instead of an unun is created.

*Figure 11-8   Photo shows the implementation of the circuit of Figure 11-7
to create a 1:2.25 balun.*

interconnection scheme is the same as the unun version. A photo of a 1:2.25
transformer constructed in this manner is presented in Figure 11-8.

11.7   Calculation of Impedance Transformation Ratios

An analysis of voltages provides a further illustration of the operation of fractional
ratio equal delay transformers. Figure 11-9 shows the three lines connected as a
1:2.25 balun, but with the outputs (high impedance side) of the lines terminated in
their characteristic impedance. At the input side, the 2 V RF source places 2 V
across the conductors of line 1. In parallel with line 1 is the series connection of
lines 2 and 3. The voltage divider comprising lines 2 and 3 divides the 2 V source
equally, placing 1 V across each of those lines.

When the transformer is operating properly, each line delivers the input voltage into a load equivalent to its characteristic impedance. In the case of the 1:2.25
transformer, line 1 delivers 2 V into 75 Ω, while lines 2 and 3 are connected in
parallel, effectively delivering their 1 V into the 37.5 Ω that results from paralleled
75 Ω loads. Line 1 is connected in series with 37.5 Ω. The final output is the sum of
the voltages and impedances: 3 V across 112.5 Ω. The validity of this analysis is
confirmed by the equivalence of the 2 to 3 V transformation and the 50 to 112.5 Ω
impedance transformation.

Knowing that our topology rule of "mirror image" series and parallel connections results in correct source and termination impedances, we can develop a method
for calculating the impedance transformation of a specific transformer, or we can
determine the transformer configuration for a desired impedance transformation.
The calculations are simplified greatly by treating each line as an ideal element, a
purely resistive source or load with a value equal to its characteristic impedance.

Let us begin with the simpler method of determining the impedance transformation of a specific transformer. As an example, we will use a five-line equal delay
transformer. The connection scheme is as follows:

End 1: Line 1 is connected in parallel with the series combination of lines 2–5.
End 2: Following the alternate connection rule, line 1 is connected in series
with the parallel combination of lines 2–5.

For convenience, a standard diagram can help describe the various connections
of transmission lines in this family of transformers. One way to annotate the

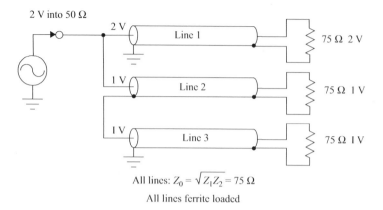

All lines: $Z_0 = \sqrt{Z_1 Z_2} = 75\ \Omega$
All lines ferrite loaded

Figure 11-9   Schematic shows a 50 to 112.5 Ω (1:2.25 ratio) three-line equal delay transformer showing the voltage distribution.

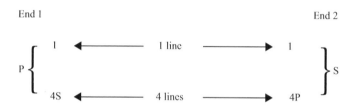

NOTE: A transformer is completely defined by one end connection—the other end must always have an opposite combination of series and parallel connections.

Figure 11-10   Notation scheme used to describe the connections for a specific five-line equal delay transformer.

connections for the example transformer is adapted from McClure's notation scheme and shown in Figure 11-10 [5]. At end 1 (left side), the connection diagram indicates that the transformer has a single line (1) and a series combination of four lines (4S). These two groups are connected in parallel (P). As required, the opposite connection scheme is indicated for end 2. Although a single end will define the entire transformer, it is helpful to include both end connections in the diagram as a reminder of the proper connection scheme. A wiring diagram for this transformer is shown in Figure 11-11.

We can determine Z_{in} and Z_{out} by simple series and parallel calculations:

$$Z_{in} = Z_0 \| (Z_0 + Z_0 + Z_0 + Z_0) = 0.8 Z_0 \tag{11-1}$$

$$Z_{out} = Z_0 + (Z_0 \| Z_0 \| Z_0 \| Z_0) = 1.25 Z_0 \tag{11-2}$$

The impedance ratio can now be calculated as

$$Z_{out}/Z_{in} = 1.25/0.8 = 1.56 \tag{11-3}$$

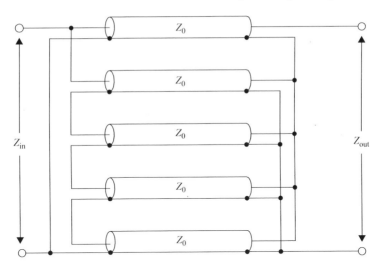

Figure 11-11   Pictorial showing the transformer described in Figure 11-10.

We can also determine the transformer configuration from the desired impedance transformation. The method used is fractional expansion, which reduces the desired *n/m* ratio into a series of fractions that define the transformer connections. Beginning with the first fraction, each fraction in the series represents alternating parallel and series connections.

Let us use the previous transformation ratio as an example. A 1.56 impedance transformation is achieved with an *n/m* ratio of 5/4. A fractional expansion is

$$5/4 = 1 + 1/4$$

This tells us that we need one line in parallel with a group of four lines in series to achieve the impedance ratio of $(5/4)^2$, which is what the example transformer will provide.

With larger values of *n* and *m*, the method of fractional expansion can provide more than one result. Each will be a valid result, but the number of lines required to construct the transformer may be different. Further description of mathematical synthesis of these transformers using fractional expansion will not be presented here. Readers interested in pursuing this topic are directed to [3–5].

11.8   Achievable Transformation Ratios

Most practical equal delay transformers will have a small number of lines. While there is no performance advantage with a simpler transformer, ease of construction, complexity, size, and weight are significant design considerations. It is possible to synthesize a transformer's design using the aforementioned fractional expansion, but it is much easier to refer to a table of transformation ratios using a small number of transmission lines. The list included in Figure 11-12 shows the transformation

| \[3-Line Transformers header] | | | |

3-Line Transformers

n/m ratio	Impedance ratio	Connection diagram End 1	End 2
3/2	2.25	P { 1 / 2S	1 / 2P } S
3/1	9	3P	3S

4-Line Transformers

n/m ratio	Impedance ratio	Connection diagram End 1	End 2
4/3	1.78	P { 1 / 3S	1 / 3P } S
5/3	2.78	P { S / 1 { 2P / 1	2S / 1 } P / 1 } S
5/2	6.25	P { 2P / 2S	2S / 2P } S
4/1	16	4P	4S

Figure 11-12   Transformation ratios and connection diagrams used for equal delay transformers with three and four transmission lines.

ratios available using three and four lines. Figure 11-13 continues the list for five-line transformers.

A wide range of transformation ratios is available using these design options. The majority of practical applications can be addressed using the transformer configurations on the lists in Figure 11-12 and Figure 11-13. If a 1:2 impedance transformation is desired, a five-line transformer can be configured for 1:1.96 ratio. For less critical needs, a three-line transformer with a 1:2.25 ratio may be quite acceptable to provide a 1:2 transformation ratio.

Similarly, ratios on the lists in Figure 11-12 and Figure 11-13 can accommodate matching between common system impedances of 50:75 Ω (1.44 or 1.56 ratio), 25:50 and 50:95 Ω (1.96 ratio), 25:75 Ω (3.06 ratio), or 50:300 Ω (6.25 ratio). While not always precise, the closest ratio often will provide matching with acceptably low VSWR and loss.

5-Line transformers			
n/m ratio	Impedance ratio	Connection diagram End 1	End 2
7/6	1.36	$P\left\{ \begin{matrix} S\left\{ \begin{matrix} 2P \\ 1 \end{matrix} \right. \\ 2S \end{matrix} \right.$	$\left. \begin{matrix} 2S \\ 1 \end{matrix} \right\} \begin{matrix} P \\ 2P \end{matrix} \Big\} S$
6/5	1.44	$S\left\{ \begin{matrix} 3P \\ 2P \end{matrix} \right.$	$\left. \begin{matrix} 3S \\ 2S \end{matrix} \right\} P$
5/4	1.56	$P\left\{ \begin{matrix} 1 \\ 4S \end{matrix} \right.$	$\left. \begin{matrix} 1 \\ 4P \end{matrix} \right\} S$
7/5	1.96	$P\left\{ \begin{matrix} S\left\{ \begin{matrix} 2S \\ 2P \end{matrix} \right. \\ 1 \end{matrix} \right.$	$\left. \begin{matrix} 2P \\ 2S \end{matrix} \right\} P \atop 1 \Big\} S$
8/5	2.56	$P\left\{ \begin{matrix} S\left\{ \begin{matrix} P\left\{ \begin{matrix} 2S \\ 1 \end{matrix} \right. \\ 1 \end{matrix} \right. \\ 1 \end{matrix} \right.$	$\left. \begin{matrix} 2P \\ 1 \end{matrix} \right\} S \atop 1 \Big\} P \atop 1 \Big\} S$
7/4	3.06	$P\left\{ \begin{matrix} S\left\{ \begin{matrix} 3P \\ 1 \end{matrix} \right. \\ 1 \end{matrix} \right.$	$\left. \begin{matrix} 3S \\ 1 \end{matrix} \right\} P \atop 1 \Big\} S$
7/3	5.44	$P\left\{ \begin{matrix} 2P \\ 2S \end{matrix} \right.$	$\left. \begin{matrix} 2S \\ 2P \end{matrix} \right\} S$
8/3	7.11	$P\left\{ \begin{matrix} S\left\{ \begin{matrix} 2P \\ 1 \end{matrix} \right. \\ 2P \end{matrix} \right.$	$\left. \begin{matrix} 2S \\ 1 \end{matrix} \right\} P \atop 2S \Big\} S$
7/2	12.25	$P\left\{ \begin{matrix} 3P \\ 2S \end{matrix} \right.$	$\left. \begin{matrix} 3S \\ 2P \end{matrix} \right\} S$
5/1	25	5P	5S

Figure 11-13   *Transformation ratios and connection diagrams used for equal delay transformers with five transmission lines.*

One realistic scenario for selecting a transformer involves a 1/4 wavelength vertical antenna over an excellent ground system. Over "perfect" ground, this vertical has a feedpoint impedance at resonance of about 36 Ω. If the vertical includes traps or multiple parallel elements covering several frequencies, it will be necessary to transform that impedance to a common feedline impedance over a significant bandwidth. A transmission line transformer is an obvious choice for the task.

To match 36 Ω to a 50 Ω feed line, the five-line transformer with a 1:1.36 transformation ratio is an outstanding choice. Surplus CATV 75 Ω cable is often accessible at no cost, and matching to that impedance would require a 1:2 ratio, with a five-line design (1:1.96 ratio) or possibly with three lines (1:2.25).

An ambitious radio amateur may want to feed two of these verticals in a broadside pattern to obtain gain. Fed in phase, the two verticals will not be far from the 36 Ω impedance if spaced 1/2 wavelength or more. Its this case, the feedpoint impedance can be carried to a junction using a 37.5 Ω cable comprising two equal-length 75 Ω cables connected in parallel. At the junction, these two lines are connected, which results in an impedance of about 18 Ω. Transforming this impedance to 50 Ω requires a 1:2.78 ratio, which is readily obtained with a four-line transformer having exactly that 2.78 ratio.

Other possible uses include matching stacked Yagi antennas to a 50 or 75 Ω feed line, transforming a 50 Ω impedance to 75 Ω to use surplus CATV hardline, or matching a feed line to the driven element in a multiband Yagi antenna.

11.9   A 50–75 Ω Equal Delay Unun Example

To test the design concepts presented in this chapter (and to get a useful accessory for Breed's ham station), a transformer for matching 50 Ω to 75 Ω was selected for design, construction and testing.

Among the ratios listed in Figure 11-12 and Figure 11-13, the ratios of 1.44 and 1.56 both have a 4% deviation from an exact 50:75 Ω (1:1.5) ratio. The 1.44 ratio was chosen because the actual characteristic impedance of coax cables ranges from 50 to 53 Ω for "50 Ω" cables, and from 72 to 75 Ω for "75 Ω" cables. With this range of impedances, the required transformation lies between 1.35 and 1.5 and will rarely be above 1.5. The transformer diagram is shown in Figure 11-14. The five lines use the following connection scheme:

$$S \begin{Bmatrix} 3P & 3S \\ 2P & 2S \end{Bmatrix} P$$

For maximum bandwidth and lowest loss, the characteristic impedance of each line must be $\sqrt{50 \times 75}$ Ω, or 61.2 Ω. This cable can be obtained by modifying RG-142 PTFE dielectric coax cable. The jacket and shield braid are carefully removed. Using the information provided by Gunston and the WinLine transmission line analysis program we can determine that increasing the inside diameter of the outer conductor by 0.0375 in will change the characteristic impedance from 50 to 61.2 Ω [6,7].

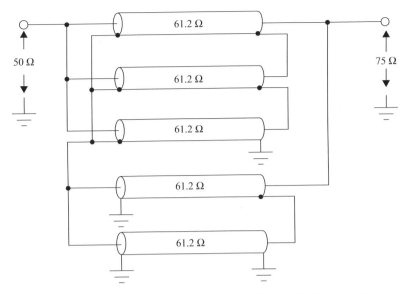

Figure 11-14 Diagram showing the connections for a 1:1.44 equal delay unun transformer used for matching 50 to 75 Ω.

This calculation can also be made using a scientific calculator. The impedance of a coax line is proportional to $\ln(b/d)$, where b is the inside diameter of the outer conductor, and d is the diameter of the inner conductor. Therefore, to get 61.2 Ω impedance using the same inner conductor of a 50 Ω line, we make the following calculation:

$$(61.2/50) \times \ln(b/a)[50\ \Omega] = \ln(b/a)[61.2\ \Omega]$$

After solving this for the new 61.2 Ω b/a value, a little algebra tells us that the modified cable needs to increase the outer:inner diameter ratio by a factor of 1.407. An additional 0.0375 in layer of dielectric material will accomplish this increase for RG-142.

A layer of heat shrink tubing was applied over the original inner conductor and its dielectric. When shrunk, 3/16 in tubing provides approximately the right additional thickness. The braid was then compressed and slid over the new inner conductor assembly, and another layer of heat shrink tubing was applied as a new jacket. The dielectric constant of the polyolefin heat shrink tubing is similar to PTFE, perhaps a bit higher in value. A quick calculation of possible errors in thickness and dielectric properties suggests that the new cable is within 5% of the desired 61.2 Ω. Figure 11-15 illustrates the modified coax lines.

The next task is to determine the necessary ferrite loading for isolation between the ends of each transmission line. Since this is an equal delay unun, the bottom line of Figure 11-14 needs no isolation since it experiences no voltage drop on the outer conductor. The third and fourth lines (from top down) experience the greatest voltage drop and were loaded with Fair-Rite 77 material ($\mu_0 = 2000$).

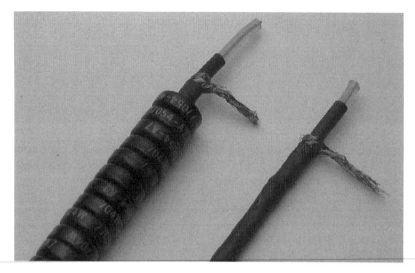

Figure 11-15   Photo showing RG-142 cable modified to raise the impedance from 50 to 61.2 Ω. The inner insulation is increased in diameter by adding a layer of 3/16 in shrink tubing. The braid is returned to the cable, and then the entire cable is enclosed in a final layer of heat shrink tubing.

The remaining two lines were loaded with the same amount of 43 material ($\mu_0 = 850$). Two of the completed lines shown in Figure 11-15 have ferrite material applied.

The five lines were then assembled and connected according to the diagram of Figure 11-14.

11.10   Performance of the 50–75 Ω Transformer

The transformer was assembled in an aluminum channel that could have a shielding cover placed over it for permanent use. The various crossover connections at the cable ends required care in construction to make crossover connections. Other experimenters may find it easier to construct these types of transformers using a small printed circuit board end plate to organize the ends of the coax lines before making the required interconnections.

First, the transformation accuracy was evaluated by terminating the transformer with a precision 75 Ω termination. The return loss at the 50 Ω end was measured using an Anzac RB-1-50 return loss bridge, with a Hameg spectrum analyzer and tracking generator as the signal source and detector. The return loss exceeded 30 dB (1.06:1 or better VSWR) from approximately 4 to 34 MHz, with lowest return loss (>34 dB) in the 6 to 10 MHz range.

This transformer uses linear loading instead of windings on a torioidal core, which requires high permeability ferrite materials to ensure sufficient isolating

Table 11-1 Measured Loss into a 70 Ω Load Using the 50:75 Ω Transformer

Frequency (MHz)	50 Ω input (W)	70 Ω Load		Loss (dB)
		RF Current (A)	Power (W)	
1.8	500	2.53	448	0.47
3.5	500	2.61	477	0.21
7.0	500	2.60	473	0.24
14.0	500	2.60	473	0.24
21.0	500	2.53	448	0.47
28.0	500	2.58	468	0.31

inductance. 77 material and 43 material may not be optimum choices for lowest loss. Although the transformation ratio is maintained with great accuracy over a wide bandwidth, the transformer's loss was not as good as desired.

Loss measurements and power handling evaluations were made using a high-power 70 Ω dummy load, an amateur radio transmitter, and power amplifier. Fortunately, Breed's high-power 50 Ω dummy load is constructed using seven 350 Ω noninductive resistors. Two resistors were removed, resulting in a 70 Ω load capable of handling high power.

To measure loss through the transformer, the 50 Ω input power was measured with a Bird Model 43 watt meter and compared with the power input to the dummy load. The load includes an RF ammeter, enabling power to be calculated using the simple formula $P = I^2 R$. The ammeter was first calibrated at 50 Ω by comparing it with the WM measurement. A calibration chart was made to compensate for small variations in the ammeter frequency response. Difficulty in estimating the RF ammeter reading below 0.1 A resolution limits the accuracy of this measurement method to about 8%.

Table 11-1 illustrates the measured loss at six frequencies from 1.8 to 28 MHz. The estimated accuracy of the data is 8%. The best performance is at the middle range of these frequencies, which corresponds to the range where the best return loss was obtained in the transformation accuracy test. Reflected power measurements at these same frequencies correspond to the earlier return loss tests, but at slightly higher VSWR since the dummy load is 70 rather than 75 Ω. VSWR readings were in the range of 1.1:1–1.25:1 at the test frequencies.

Transformer heating was carefully observed during the loss testing. At the 500 W power level used in the tests, heating of the ferrite material was noticeable. However, even at frequencies where the highest loss was measured, heating was modest. For example, after 30 sec of continuous operation at 500 W at 1.8 MHz, the transformer was subjectively "hot," but less than 60°C. This is a significantly smaller temperature rise than expected with 52 W of disspation, suggesting that the loss measurements may be pessimistic.

The pattern of heating was observed to change with frequency. At the lower frequencies, the center ferrite cores were the warmest, while at the higher frequencies the heating was more even across the length of ferrite material.

An unexpected result was that the temperature rise at any given point along the transformer was lower at high frequencies since the dissipation was not concentrated at one location.

High-power testing also included brief operation at 1000 and 1500 W. As expected, the temperature rise was greater, and operation for an extended period of time at this power level is not recommended. However, the RG-142 coax cable, aided by the distribution of power among the windings, is within its rating at this power level.

This example demonstrates the ability of equal delay transformers to provide accurate fractional ratio impedance transformations. In practice, improved loss performance may be obtained with an alternate construction method, such as several turns of coax wound on toroidal cores of lower loss ferrite material. However, the example transformer is useful in 50 to 75 Ω impedance transformation applications at power levels up to a few hundred watts.

References

[1]   Lewis, W. A., "Low-Impedance Broadband Transformer Techniques in the HF and VHF Range," Working paper, WP-8088, Collins Radio, July 1995.

[2]   Blocksome, R. K., "Practical Wideband RF Power Transformers, Combiners, and Splitters," *Proceedings RF Technology Expo 86*, Cardiff Publishing Co., 1986, pp. 207–227.

[3]   Sabin, W. E., and Schoenike, E. O., eds., *HF Radio Systems & Circuits*, rev. 2nd ed., Noble Publishing Corp., 1998, Ch. 12, "Solid-State Power Amplifiers," R. K. Blocksome, author.

[4]   McClure, D. A., "Broadband Transmission Line Transformer Family Matches a Wide Range of Impedances," *RF Design*, Feb. 1994, pp. 62–66.

[5]   McClure, D. A., "Broadband Transmission Line Transformer Family Matches a Wide Range of Impedances—Part 2," *RF Design*, May 1995, pp. 40–49.

[6]   Gunston, M. A. R., *Microwave Transmission Line Impedance Data*, Noble Publishing Corp., 1997, Ch. 2.

[7]   WinLine software, Noble Publishing Corp., Norcross, GA, 1995, information available at http://www.noblepub.com

Chapter 12

Simple Test Equipment

12.1   Introduction

The world has changed dramatically since Sevick wrote this chapter in the 1980s. At that time, your options for impedance measurement would be to find someone with access to very expensive commercial equipment or to build your own. RF tools were either laboratory grade or TV repair grade; experimenter-grade RF tools did not emerge until MFJ Enterprises was established. Digital and computer tools were not available. Personal computers were in their infancy, with IBM, Compaq, and Apple as the major (and expensive) commercial vendors.

Now, microcomputer-based tools are readily accessible, almost matching the capabilities of 1980s commercial test tools and available at prices within most experimenters' budgets. The commercial equipment that Sevick used was likely Hewlett-Packard (now Agilent) or Tektronix vector network analyzers (VNAs) costing tens (or even hundreds) of thousands of dollars. If your budget does not allow you to purchase experimenter-grade test equipment, Sevick's original designs are reproduced starting in section 12.4. However, unless your lab contains a significant number of required parts, it is unlikely that you will be able to produce your own test equipment for a cost much below that of modern experimenter-grade tools.

12.2   Transformer Impedance Measurement

Many companies manufacture test equipment which is essentially a single-port VNA in the price range of $200–$500. VNAs allow measurement of parameters including resistance and reactance. Most are intended as antenna analyzers or SWR analyzers, but many also include additional functions such as frequency counting and time domain reflectometry. These devices include a stable, constant amplitude (more or less) oscillator, a frequency counter (for accurate frequency readout), and a Wheatstone bridge. Manufacturers include MFJ Enterprises, AEA, Autek, and RigExpert. The American QRP Club also sells a kit for approximately $300 for a DDS-controlled version that covers 1 to 60 MHz. An Internet search will likely also reveal others. The time domain reflectometers (TDR) of devices in this price range are often frequency domain reflectometers that use the inverse Fourier transform to convert a frequency scan into a time domain representation. In general they assume 50 Ω characteristic impedance.

Table 12-1   Attenuation versus Load Impedance for
Attenuators Used as Precision Loads

Attenuation (dB)	Load
1	436
2	221
3	151
4	116
5	96
6	83
3 + 3	75.5
10	62.3
15	53.5
30	50
6 + 6	41.5
3 + 30	37.6
30 + 30	25
30 + 10	27.7

The commercial equipment Sevick used at Bell Labs was capable of both one- and two-port VNA impedance measurement. A two-port device will measure transmission through a device (e.g., the power loss through a transformer). These typically also measure S-parameters of a network. A few experimenter-level two-port VNA devices, such as two produced by Mini Radio Solutions, are available in the $500 range, and they will cover our frequencies of interest.

Measuring impedance transformation at higher frequencies requires a load with minimal inductance or capacitance. Resistors with leads will always have some amount of inductive reactance, especially above 50 MHz. A better alternative is to use a BNC T connector and combine attenuators in parallel to achieve the desired resistance. Attenuators with 5% accuracy are available from Emerson Connectivity Solutions (through distributors) or Mini-Circuits (direct) in the price range of $7–$10 per attenuator. It is possible to generate a fairly large range of load resistances with a small selection of attenuators. Table 12-1 provides a representative set of load values that are possible using unterminated attenuators alone or in parallel.

12.3   Transmission Line Impedance Measurement

Another task we encounter is measuring the characteristic impedance of the transmission lines for our transformers. We need to determine if we have created a line with the optimum characteristic impedance. We might be inclined to employ the single-port VNA tool for the job, but we would be disappointed because they make several assumptions regarding the load. The principal assumption is a 50 Ω transmission line. Other impedances would require knowing the exact length of the line as well as the velocity factor of the line. Since we generally do not know these, we need a different method. TDR will allow us to measure characteristic impedance of even very short lines. It is possible to build a useful TDR from a fast rise time pulse generator and a fast oscilloscope.

Fortunately, both stand-alone and PC-based digital storage oscilloscopes are also within the price range of most experimenters: between $350 and $1300 (manufactured by Rigol, Agilent, and Textronix). It is also possible to use a very fast analog scope if the pulse generator operates at a high frequency. A scope's main requirement for measuring short lines is a horizontal time per division of 2–10 ns/div.

Figure 12-1 shows the test setup necessary to create a TDR. Scopes typically have 5–20 pF capacitance in parallel with the 1 MΩ input. This capacitance creates a problem with the fast rise time used in a TDR. You will see ringing for a short time at the rising edge of the pulse. Figure 12-2 shows the results of a simple TDR using a TDS2012 oscilloscope for display and a HP32120A arbitrary waveform generator (sync output) for pulse generation. The line being measured is 100 Ω twisted pair from CAT-6 cable with 150 Ω BNC attenuator termination. The ringing is not too bad, but it adds some uncertainty to the measurement.

A better pulse generator is shown in Figure 12-3. An ordinary DIP clock oscillator is passed to a 74HCT14 Schmidt trigger. The gate creates a very fast rise and fall time (typically less than 2 ns) to present to the cable under test.

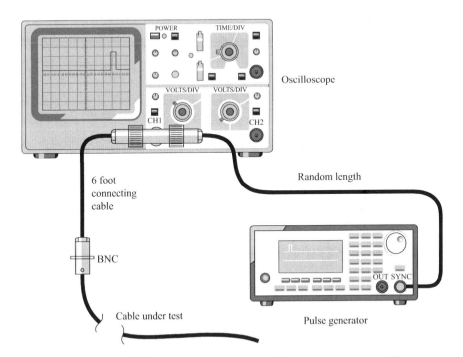

Figure 12-1 Pictorial shows a setup to use a pulse generator, an oscilloscope, and a known length of 50 Ω cable to create a time domain reflectometer for transmission line impedance measurement.

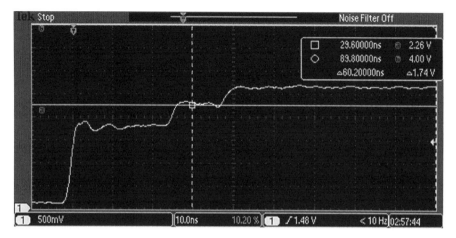

Figure 12-2   Screen capture from an oscilloscope set up as a TDR. The unknown line is a 24 in length of CAT-6 twisted pair line.

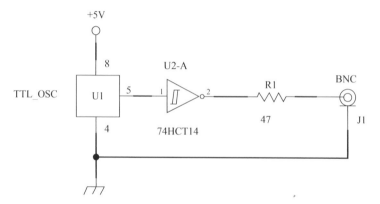

Figure 12-3   Schematic shows a simple circuit using a clock oscillator and a Schmidt trigger to create a fast rise time pulse generator for a TDR source.

The 47 Ω resistor creates a nominal 50 Ω source resistance for use in measuring the unknown value.

A short piece of 50 Ω cable separates the pulse generator and the unknown line. The initial pulse will travel down the 50 Ω line for a short time until it reaches the unknown line. Once the pulse hits the line with an impedance other than 50 Ω, a return pulse will be generated; an impedance higher than 50 will cause a higher voltage to be reflected and a lower impedance will cause a lower voltage to be reflected. We can predict the actual voltage on the line for the return step using the voltage divider equation:

$$V = V_{in}\left(\frac{Z_l}{Z_S + Z_l}\right)$$ (12-1)

where

Z_l = the load impedance
Z_S = the source impedance
V = the measured voltage
V_{in} = the open circuit voltage of the generator

The Z_S transform of the equation is used to calibrate the pulse generator impedance:

$$Z_S = \frac{50}{V_{50}} * (V_{in} - V_{50}) \tag{12-2}$$

where

V_{50} = the voltage across the 50 Ω calibration load

The Z_L transform of the equation is used to determine the unknown impedance:

$$Z_L = \frac{Z_S * V_{Z_L}}{(V_{in} - V_{Z_L})} \tag{12-3}$$

where

V_{Z_L} = the voltage measured across the unknown load

Using the TDR requires us to calibrate the display. First the scope is connected directly to the pulse generator output to measure the open circuit voltage (V_{in}). The next step is to place a reference 50 Ω load on the TDR. This voltage in equation (12-2) is then utilized to determine the true source impedance of the pulse generator (Z_S). The scope trace in Figure 12-2 shows the "B" cursor value of 4.0 V, which was the open circuit voltage measured before the screen shot. My setup measured Z_S as 77 Ω. From the cursor in Figure 12-2 we measure 2.26 V for our CAT-6 cable. Equation 12-3 shows that our cable is 101 Ω. Since CAT-6 cable is designed to be 100 Ω, we see that our test setup is very accurate!

It is important to terminate the transmission line under test with a resistance close to its characteristic impedance. Failure to include this step will result in reflections that will distort the display. If you happen to terminate the line in its characteristic impedance, you will not see a bump in the display corresponding to the end of the line; the pulse will continue without reflecting back. Of course, that is very unlikely to be the case. If I had a 5 dB attenuator available for my CAT-6 experiment, I would have seen only a very small dip in voltage when the pulse reached the end of the line. The precision termination also provides a secondary calibration point to verify that your line is the intended impedance.

12.4   Home-Built Equipment

The remainder of this chapter is directed to persons who do not have access to sophisticated test equipment and must rely on simple equipment that can be constructed from readily available parts. We cover homemade test gear that can give surprisingly good results. Measurements Sevick made on this equipment matched very closely with high-precision laboratory test sets [1–3].

When Sevick designed the following equipment, mechanical and analog functions for frequency generation and readout were generally available and usually at reasonable prices at most metropolitan-area radio or electronic stores. The world has changed such that mechanical readouts such as vernier drives and dials are no longer manufactured and local parts stores are nonexistent. The parts in the following designs are quite old, but most can still be purchased from sources such as RF Parts, Mouser, and DigiKey. Planetary reduction drives are still manufactured by Oren Elliot Products, but you will need a means for frequency readout. Fortunately, frequency counters are very inexpensive compared with 1980s prices. Kits or inexpensive commercial frequency counters (under $50) can be found by searching the Internet.

All of the important transformer characteristics can be measured using these basic tools. Parameters such as transformation ratios, high and low frequency performance, optimum impedance levels for the types of windings used, and the characteristic impedances of bifilar windings can be readily determined. Each parameter can be evaluated with a simple resistive bridge and a general-coverage signal source that uses junction field-effect transistors (JFETs). That elusive parameter known as efficiency can also be obtained satisfactorily by a simple comparative technique described in section 12.8. This is an indirect measure of the in-band loss (only 0.02–0.04 dB). (Direct measurements can be made only with highly complex laboratory apparatuses.) As an added bonus, this simple equipment can be effective for measuring the important parameters of vertical antennas, that is, the resonant frequency and the resistance at resonance.

12.5   The Wheatstone Bridge

The simple resistive bridge, known as the Wheatstone bridge, is shown in Figure 12-4a. It is balanced when no voltage exists between terminals D and B.

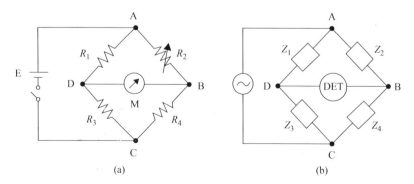

Figure 12-4   (a) DC Wheatstone bridge. (b) AC equivalent to the Wheatstone bridge.

At this point of balance, no current flows through the galvanometer (M) and a null is indicated. This occurs when the ratios in the arms are

$$R_1/R_3 = R_2/R_4 \qquad (12\text{-}4)$$

If R_4 is the unknown resistance, then equation (12-4) can be rewritten as

$$R_4 = R_2 \times R_3/R_1 \qquad (12\text{-}5)$$

In general, the ratio R_3/R_1 is some conveniently fixed value such as 1, 2, or 3, and R_2 is made variable for balancing purposes. By using many known values of R_4, the dial of the variable resistance (R_2) can be readily calibrated. For example, if the dial has 100 divisions to cover one complete rotation of R_2 (usually 330°), and $R_1 = R_3$, then the dial can be read directly in ohms if R_2 is a 100 Ω potentiometer. If $R_3/R_1 = 2$, then each division equals 2 Ω (or 200 Ω full scale). Many other combinations are possible. The simple AC impedance bridge shown in Figure 12-4b is an extension of the Wheatstone bridge. It is balanced when the voltage between terminals D and B is zero. This occurs when

$$Z_1/Z_3 = Z_2/Z_4 \qquad (12\text{-}6)$$

The unknown impedance (Z_4) becomes

$$Z_4 = Z_2 \times Z_3/Z_1 \qquad (12\text{-}7)$$

If Z_3 and Z_1 are pure resistances, and Z_4 is complex (i.e., $R_4 + jX_4$), then Z_2 must have a resistive and a reactive component for balance. Again, other combinations are possible. AC bridges can be used to trade capacitive reactance for inductive reactance by properly placing the arms in the bridge.

12.6   A High-Frequency Resistive Bridge

Accurate AC bridges are capable of measuring reactance as well as resistance but are rather difficult to construct and calibrate. Surprisingly, the simple resistive bridge, when carefully built to reduce parasitic elements, can measure broadband transformer performance. The bridge is effective up to 100 MHz or more. The key is to understand the nature of the null. With pure resistive loads (R_4), the nulls are very sharp and deep. This is the standard indication for any purely resistive load. A load that includes reactance will have a less pronounced null. Within a transformer passband, its input impedance is resistive when it is terminated with resistive loads, so a sharp null can be expected. To reduce the effect of parasitic inductance, it is advantageous to terminate the transformer in question with a resistive load on its high-impedance side and measure from the low-impedance side. Transformation ratios are readily measured within the passband of the transformers because reactive components are not created. The devices act as ideal transformers.

The low frequency response can be determined with the aid of a variable frequency signal source. The input impedance of the terminated transformer takes on an inductive component at the low frequency end where the choking action of the

coiled transmission line becomes inadequate, causing the null to be less pronounced. By lowering the frequency of the signal source to where the meter reading
at the null increases noticeably (usually 10–20% of full scale), a fairly good measure of the low frequency capability is obtained.

In a similar fashion, the high frequency response can be determined by increasing
the frequency of the signal source until a noticeable increase in the null reading takes
place. This indicates a reduction of the transformer's capability at the high frequency
end. Also of interest is that the optimum impedance level for maximum high frequency response can be determined easily. By varying the value of the terminating
resistance and noticing the depth of the null at the high frequency end, the best
impedance level for the type of winding used on the transformer can be determined.
As will be shown in section 12.6, even the characteristic impedance of short (coiled or
straight) transmission lines can be evaluated quite accurately by this method.

Figure 12-5 shows a circuit diagram for a linear resistive bridge. A germanium
or Schottky diode is used for D1 because of its lower forward voltage drop compared with a silicon diode. The fixed 33 Ω resistors in the top arms of the bridge are
not especially critical. Values of 47 to 68 Ω have also been used. Importantly, the
resistors should be closely matched and also noninductive. The potentiometer
should also be noninductive and either carbon composition or Cermet rather than
wire wound. An inexpensive 0–100 dial can be placed under the knob for the
potentiometer. To calibrate the dial, various calibration resistors can be connected
from J2 (via a banana plug) to the grounded binding post. You should perform the

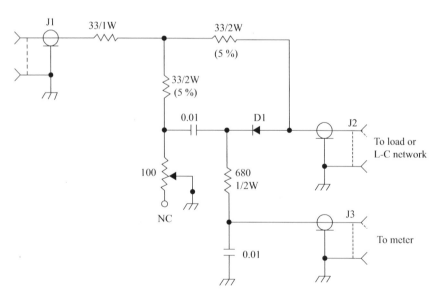

*Figure 12-5   Schematic shows a practical AC impedance bridge. The input is from
a variable frequency sine wave generator, and the output is a DC
current that corresponds to the amount of imbalance of the bridge.*

calibration with 1% noninductive resistors (which are available from distributors such as Mouser and Digi-Key), and at 3.5 MHz or lower in frequency. If you are buying attenuators for characteristic impedance measurements, then those are also good calibration tools. In this frequency range, the inductive effect of the resistor leads is minimized. This bridge, if built with short leads, works well up to at least 100 MHz, with an accuracy of about 1 Ω over its linear scale.

Figure 12-6 shows a circuit diagram for the current amplifier used in conjunction with the resistive bridge of Figure 12-5. With a 50 μA panel meter, the maximum sensitivity is 7 nA full scale. Analog panel meters are no longer inexpensive or readily available. It is probably cheaper to use an inexpensive digital voltmeter than an analog panel meter for the readout and connecting the meter to points A and B. Radio Shack carries several meters for less than $30. A digital meter should be set to a fixed range rather than letting it auto-range so that the null is easier to see. Other inexpensive sources for meters can be found on the Internet. An inexpensive analog volt-ohm meter may be an even better choice for observing the dip in the response since the dip is more easily perceived than watching numbers change. Several interesting features should be noted in Figure 12-6:

1. The circuit is a noninverting amplifier. The input impedance of the op-amp is thousands of times higher than the 10 k resistance, so performance depends solely on the 10 k input resistor.

2. Recommended op-amps are TL092, MAX407, or LTC6078. These are low input bias amplifiers that also give rail-to-rail operation. The second amplifier creates a low impedance ground reference that is 1 V above the negative connector of the battery. This gives 8 V maximum output.

3. The 100 kΩ resistor can be switched in parallel with the 5 MΩ potentiometer to provide a high/low sensitivity control.

This instrument is particularly useful for measuring vertical antennas.

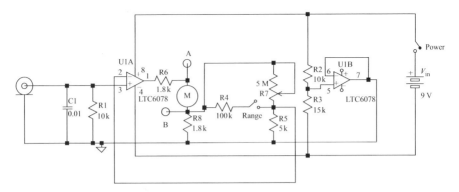

Figure 12-6   Schematic shows a practical current amplifier using a modern dual op-amp IC. Suggested ICs are low bias circuits from Maxim, Linear Technology, or TI. A VOM or DVM can be inserted at terminals A and B instead of using an analog panel meter.

12.7   Signal Generators

When you measure the performance of broadband transformers (as well as vertical antennas), it is convenient to have signal sources that are continuously variable over a very large frequency range, compatible in signal level with the impedance bridges, constant in output, and portable. The sources in this section fulfill these requirements.

Figure 12-7 is a schematic using JFETs in a Colpitts oscillator and a source follower. The transistors, Q1 and Q2, are J310 and J309, acquired from DigiKey or RF Parts. Both components are capable of operating beyond 100 MHz. J309 is particularly useful; its high transconductance of 20,000 μS gives an output of 50 Ω as a source follower. Its input impedance is also extremely high, making it ideal for use in a decoupling stage. Motorola's (now Fairchild Semiconductor) MPF102 and Texas Instruments 2N5397 also have worked satisfactorily in the circuit of the figure. Sevick used the following inductors from the J.W. Miller Co: L1 is a 4508 (24–35 μH); L2 is a 4503 (1.6–2.8 μH); and L3 is a 4501 (0.4–0.8 μH). J. W. Miller no longer manufactures those part numbers, but close values from Toko America can be found at DigiKey. Toko parts are TK2410 (27 μH), TK3146 (1.5 μH), and TK3011 (0.6 μH). The adjustable feature is not required since you will use a frequency counter for frequency display. The only requirement is that the inductor chosen gives tuning that overlaps across ranges. The tuning capacitor Sevick used was a three-gang 350 pF broadcast tuning capacitor. The 700 pF value is obtained by using two sections in parallel.

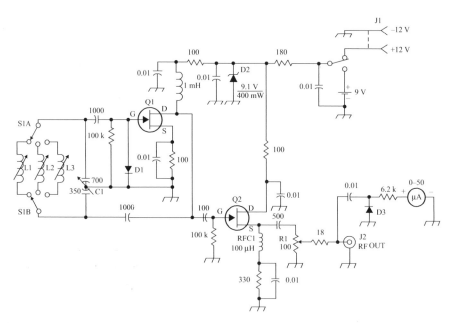

Figure 12-7   Schematic shows a Colpitts oscillator that generates a reasonably clean and stable sine wave.

Another generator using a Hartley oscillator and source follower is shown in Figure 12-8. The differences between the signal sources of Figure 12-7 and Figure 12-8 are in the sizes of the variable capacitors and the number of inductors. Figure 12-8 uses a 365 pF dual gang variable instead of a 350 pF triple gang variable. The inductors used in the unit of Figure 12-8 are also from the J.W. Miller Co. or Toko America and have the following numbers and values:

L1: 4409 (68–130 µH) Toko TK2417
L2: 4408 (30–69 µH) Toko TK2410
L3: 4506 (9–16 µH) Toko TK2429
L4: 4503 (1.6–2.8 µH) Toko TK3110
L5: 4502 (1.0–1.6 µH) Toko TK3146
L6: 4501 (0.4–0.8 µH) Toko TK3011

This signal source has a broader range; it operates from 1 to about 70 MHz with practically constant output. By removing half of the rotor plates of the variable capacitor in Figure 12-7, the high frequency range is increased to about 100 MHz. This procedure tends to spread out the tuning capability in the lower ranges, which is helpful in more accurately determining the frequency.

Surprisingly, eBay and Amazon are reasonable places to search for test equipment. I found the Lodestar (Taiwan) RF signal generator on Amazon for $165. This generator has a vernier dial just as Sevick used in the 1980s and looks

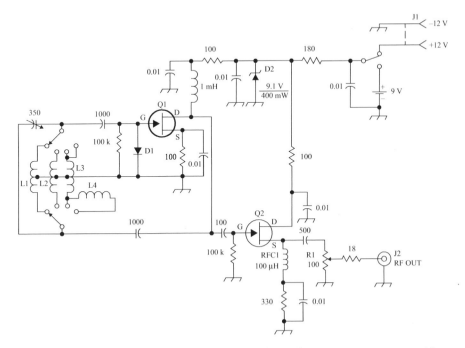

Figure 12-8   Schematic shows a Hartley oscillator that generates a reasonably clean and stable sine wave.

like his design may have been used as an inspiration. I would be hard pressed to build a generator with the mechanical and electronic components for only $165, not to mention the time required to build the equipment. Numerous companies from China and Taiwan supply kits based on the Analog Devices AD9850 DDS integrated circuit. This IC will produce a sine wave output from below 1 Hz to hundreds of MHz, so these kits are an alternative if you choose to build your own generator. The boards are generally in the $10 range but require a computer of some sort to set the frequency.

The home-built RF sources designed by Sevick have more than adequate stability from 1 to 30 MHz. This is because the bandwidths of transmission line transformers and ground-fed antennas (even on 160 m) are relatively wide, and therefore stability is not an issue. Homemade coils using small plastic forms for the higher frequencies and small powdered iron toroids (red mixture, $\mu = 10$) for the lower frequencies are appropriate. Since these coils are not variable (they lack the adjustable powdered iron slug), careful pruning should bring them into the proper ranges.

12.8   Efficiency Measurements—The Soak Test

As mentioned in previous chapters, if proper ferrite materials are used for the cores of transmission line transformers, they can exhibit outstandingly high efficiencies. This occurs because of the canceling effect of the transmission line currents. Energy is transmitted by transmission line mode instead of by flux linkage as in a conventional transformer. This holds true at the transformer's high frequency limits, where standing waves come into play to create a complex transformation ratio possibly different from that of the mid-band ratio.

If an accurate gain and phase test set for determining efficiency is not accessible, a simple technique Sevick called the soak test can be employed with surprisingly good results. It is able to distinguish transformers that are 95% efficient or less from those that are 98% or 99% efficient. The method involves using a known, efficient transformer in series with an unknown transformer. The transformers are connected back-to-back and inserted in a coaxial cable line where appreciable power is transmitted. For example, a 4:1 step-down transformer (e.g., 50:12.5 Ω) would be in series with a 1:4 step-up transformer. A 50:200 Ω transformer could equally be in series with a 200:50 Ω transformer to obtain the original impedance of the coaxial cable. The same is true for baluns, fractional ratio transformers, and other devices with higher transformation ratios.

The soak test is a qualitative technique that can be augmented with test equipment. Sevick simply used touch as a qualitative method to identify relative temperature rise. This is a fairly capable method that I have used as a semi-conductor applications engineer:

    40°C: Lukewarm, warmer than room temperature, but barely
    50°C: Warm, noticeable heat but not uncomfortable to touch
    60°C: Hot, uncomfortable to touch, but can hold for a long time
    70°C: Very hot and cannot hold for any length of time

After power is applied for several minutes, touch the transformers (with the power off!) to see if a noticeable temperature rise has occurred. Transformers with proper ferrite cores and with no. 14 or no. 16 wire on toroids of 1 1/2 in OD or greater or on rods of 1/2 in in diameter virtually have no detectable (by touch) temperature rise while handling 1 kW of CW power. Transformers with efficiencies of 95% or less show noticeable temperature rise. Incidentally, the transformers Sevick used in his soak test were a toroid 4:1 in series with a rod 1:4. No. 18 wire was used on both cores, which were Q1 material. These transformers became only mildly warm and exhibited no permanent damage when operating at the 1 kW peak power level. The rod transformer was a little warmer than the toroid because it had 40% more wire. At 200 W, neither transformer showed any perceptible temperature rise.

A more scientific method is to actually measure the temperature rise to estimate the efficiency of your transformer. A K-type thermocouple meter is quite inexpensive. I found one new on Amazon for $35, and it included two K-type thermocouples. A thermocouple is essentially a short circuit with respect to electric fields, so placing a thermocouple directly on the core even while operating the transmitter is not likely to affect the temperature reading from induced RF. You will know immediately if RF is affecting your thermocouple readings.

The key is the formula:

$$\Delta T = (\text{Power}/\text{Area})^{0.833} \qquad (12\text{-}8)$$

We see that temperature rise is a function of the power loss in the transformer and the surface area. Surface area is very easy to calculate for both toroids and rods. You will want to use the ferrite material surface area since the copper adds only slightly to the area and conducts heat significantly better than the ferrite. Notice that we measure temperature rise rather than absolute temperature. The temperature rise is also the long-term temperature rise, so we can use the time constant equation to estimate the total temperature rise. The temperature will be 63% of max at one time constant and 86% of max at two time constants. The three points (start, 63%, and 86%) will let us find the max temperature rise.

So the method to use is to connect the two transformers back to back so that the dummy load is matched to the transmitter. We affix the thermocouple to the transformer to test. Then we start the transmitter at the power level to test and take a temperature measurement every 5 sec. Be sure to stop immediately if the temperature shoots up to 80°C in a short amount of time. You can use Excel, MATLAB, or another math simulation tool to curve fit the data to find the maximum temperature rise.

We rearrange equation (12-8) to determine power loss:

$$\text{Power} = \Delta T^{1/0.833} \times \text{Area} = \Delta T^{1.2} \times \text{Area} \qquad (12\text{-}9)$$

12.9 Characteristic Impedance Measurements

As was stated many times throughout this text, when adequate isolation (due to the coiling of transmission lines around a core or the threading through beads) exists between the input and output, these devices then transfer the energy by efficient transmission line mode. Thus, their designs mainly depend on transmission line theory and practice. As with conventional transmission lines, the characteristic impedance also plays a major role with transmission line transformers. Characteristic impedances appreciably greater or smaller than the optimum value can seriously affect the high frequency response. The design goal is to be within 10% of the optimum value. For the 1:4 transformers, whether they be Ruthroff or Guanella designs, the optimum characteristic impedance is one-half the value of the resistance on the high impedance side. For low impedance coaxial cable or rectangular line transformers, the optimum characteristic impedance has been found experimentally to be 80–90% of this value. The curves in Chapters 4 and 5 show the characteristic impedance of various kinds of transmission lines obtained with the simple resistive bridges described in section 12.4. These results were from measurements on transmission lines only 10–20 in in length! The resistive bridge, with a very sensitive detection arrangement, is an excellent detector of phase (the depth of the null is diminished when the bridge sees a nonresistive termination). Thus, terminating these short transmission lines (straight or coiled) with various non-inductive resistors until the depth of the null approaches that of a pure resistor (resulting in a "flat" line) gives a quick and accurate measurement of the characteristic impedance. With this method (which rivals that of any sophisticated impedance bridge), the accuracy is determined solely by the calibration of the resistive bridge or the true values of the terminating resistors. The only complication versus using TDR is that it is necessary to have a very large selection of 1% noninductive resistors and a lot of time to work to identify the deepest null.

The following are comments and suggestions regarding measurements of characteristic impedances (especially with the equipment described in this chapter):

1. In measuring impedances between 90 and 250 Ω, the range of the bridge in Figure 12-5 has to be increased 2 1/2 times. Sevick found that replacing the 100 Ω potentiometer with a 250 Ω potentiometer and the two 33 Ω resistors (in the top arms) with 68 Ω resistors, results in an excellent resistive bridge for this range of impedances.

2. A straight wire transmission line always yields a higher impedance value than a coiled one. This is due to the proximity effect of adjacent bifilar turns in a coiled transmission line. The increased dielectric constant of the core also reduces the characteristic impedance. The smaller the spacing is between bifilar turns, the lower the characteristic impedance.

3. A straight coaxial cable transmission line (of the low impedance types described in this book) always yields a higher value than the one that is coiled around a core. This is because the effective spacing between the inner conductor and the outer braid decreases due to bend radius with coiling. The proximity of neighboring turns does not effect the characteristic impedance but does affect the parasitic capacitance and hence the high frequency response.

4. The most accurate data were obtained between 10 and 20 MHz. In this frequency range, the phase angle can still be large (if the termination is not equal to Z_0) and

easily detected. Further, the parasitic inductance of the leads of the terminating resistors is minimal.

5. For all forms of measurements dealing with transmission line transformers, Sevick found the best settings on the current amplifier to be 75–100% of full sensitivity and on the signal source to be about 25% of full scale. At high frequencies, where considerable phase angle can exist, the sensitivity of the current amplifier has to be reduced appropriately.

6. For the most meaningful readings, the characteristic impedance measurements should be made in the transformer's final configuration. In this way, all interactions between the windings are taken into account.

References

[1] Geldart, W. J., G. D. Haynie, and R. G. Schleich, "A 50-Hz–250-Mhz Computer Operated Transmission Measuring Set," *Bell Systems Technical Journal*, Vol. 48, No. 5, May/Jun. 1969.

[2] Geldart, W. J., and G. W. Pentico, "Accuracy Verification and Inter-comparison of Computer-Operated Transmission Measuring Sets," *IEEE Transactions on Instruments and Measurement*, Vol. IM-21, No. 4, Nov. 1972, pp. 528–532.

[3] Geldart, W. J., "Improved Impedance Measuring Accuracy with Computer-Operated Transmission Measuring Sets," *IEEE Transactions on Instruments and Measurement,* Vol. IM-24, No. 4, Dec. 1975, pp. 327–331.

Chapter 13

Construction Techniques

13.1 Introduction

This chapter is concerned with the practical considerations of transformer construction: how to (1) select the proper ferrites; (2) wind rod and toroidal transformers; (3) construct low impedance coax cable; and (4) handle and take care of ferrite transformers. The techniques described in the following sections evolved over many years of winding hundreds of transformers (it took Sevick about three attempts to arrive at a final design). Certainly many other techniques can do the job as well (or even better), but the ones reviewed worked well for Sevick.

13.2 Selecting Ferrites—Substitutions

Measurements on many ferrites from major manufacturers have shown that the highest efficiencies have been obtained from nickel-zinc material with permeability less than 300. Manganese-zinc ferrites have much higher permeability but high loss when used as cores for transmission line transformers. They are not recommended in RF power applications. For those who have obtained some unknown rods or toroids and want assistance in identifying them, here are some suggestions:

1. *Appearance*: Ferrites can come in all shades of black and gray-black. They can be either shiny or dull. In some cases they have a protective coating for the wire. Therefore, low permeability nickel-zinc ferrite is indistinguishable in appearance from high permeability nickel-zinc ferrite or from manganese-zinc ferrite. Powdered iron toroids usually have distinctive protective coatings. They are not, however, recommended for RF power transformer duty because of their very low permeability. The popular T-200-2 toroid has a clear enamel-like finish and a definite coating of red. It has a permeability of only 10 and is called the "red mixture." Other coatings (e.g., yellow, blue) for low permeability toroids should be used at higher frequencies for inductors or conventional transformers.
2. *Magnet test*: All ferrites and powdered irons are attracted to magnets. Therefore, testing with a permanent magnet is useless.
3. *Electrical test*: Toroidal cores can be tested for permeability by an inductance measurement. The measurement can be made directly with an inductance

meter or indirectly by a resonant circuit. Either measurement involves the geometry of the core and the number of turns used in the winding. Rod cores, because of their large air path and high reluctance, don't lend themselves as readily to similar measurements. But all of the rods seen by Sevick at surplus houses and flea markets have been made of material with a permeability of 125 and are very likely usable. This material is used in AM radios (loop stick antennas) and is excellent for transmission line transformers.

4. *Power test*: For those who are not fortunate enough to have access to sophisticated test equipment, another avenue is still available. The simple soak test will quickly identify if a core is suitable. One can quickly find out, by the temperature rise, if the ferrite is suitable for use.

5. *Substitution*: Generally, all of the low permeability ferrites from the various manufacturers can be interchanged without any significant difference in performance. Therefore, it is a matter of selecting the right permeability range. The following is a list of codes for equivalent ferrites:

$\mu = 35$–50: 67, C2075
$\mu = 100$–175: 61, 4C4, C2050, C2025
$\mu = 250$–300: 64, 66, 52

13.3   Winding Rod Transformers

Transmission line transformers with rod cores should find many applications when matching 50 Ω coax cable to lower impedances. At these low impedance levels, the coiled transmission lines on rod cores can easily offer sufficient reactance to prevent the unwanted currents and still allow for high frequency operation. Toroidal cores, with their closed magnetic paths, require fewer turns (and hence yield higher frequency responses) but are not necessary at low impedance levels. Further, rod transformers are actually easier to wind than their toroidal counterparts.

The following describes a technique for winding a bifilar transformer such that the winding is tight and is therefore optimized for both its electrical and mechanical properties. First, a single winding (as tightly wound as possible) is placed on the rod. A second winding, held in place by soldering it to the first winding, is started on the inside of the first winding and is then squeezed (or stuffed) between the turns of the first winding. The first winding is thus expanded as if it were a spring, which shrinks the inner diameter. As the second winding is completed, the two windings are not only tight to each other (resulting in the lowest characteristic impedance possible with the wire) but are also literally held fast to the rod. If the windings are somewhat loose on the rod, then small pieces of 3M no. 27 glass tape or no. 92 polyimide tape, on both ends of the windings, will prevent the rod from falling out. If a trifilar transformer is to be wound, then the first winding should have spacing between turns of about one wire diameter. In turn, a quadrifilar transformer should have a spacing of about two wire diameters and a quintufilar transformer about three wire diameters. In each case, the succeeding windings are soldered, at the ends, with the preceding windings before they are wound on the rods.

And finally, a few words are given on tapping windings. In some cases, impedance ratios require tapping the windings. For example, a trifilar winding connected as in Figure 7-13 yields a ratio of 1:2.25. Tapping winding 5–6, close to terminal 6, can yield a broadband ratio very near 1:2. Successful taps have been made by first filing about an 1/8 in wide groove around the wire with the edge of a small, fine file. Then a copper strip or a no. 14 or no. 16 gauge wire, which has one end flattened about 3/8 in long, is wrapped around the groove and soldered. The soldered connection is then rendered smooth by the edge of the file. Two or three sections of 3M no. 92 tape are then placed over the soldered area to provide mechanical and electrical protection for the adjacent windings.

13.4 Winding Toroidal Transformers

Toroidal cores offer the greatest margin in bandwidth and power because of their closed magnetic path. This allows for the use of higher permeabilities (which rods do not allow) and much shorter windings than their rod counterparts. With ML or H Imideze wire or coax cable, toroidal transformers are considered top of the line. Since so little flux enters the core of a transmission line transformer, the objectives with toroidal transformers are to use the smallest core and the highest permeability allowed by the size of the conductors and the requirement on efficiency. In many applications, cores only 1 1/2 in OD can handle the full legal limit of power allowed for amateur radio use. Larger cores have to be used at the higher impedance levels to obtain the required choke inductance and characteristic impedance. There are two major differences in winding toroidal transformers compared with rod transformers:

1. The conductors (except for a single coax cable) are bound together and wound as a ribbon. It is not possible to accurately wind each wire individually and maintain proximity.
2. Because more than one conductor is wound at a time, and considerably more bending and unbending is experienced, work hardening of the wire is much more severe. Another issue is that the ribbon tends to twist as it is wound around the core. Care should be taken to ensure that the ribbon does not make a 180° twist as it goes through the core. Considerably more force is required in the winding process. The thumb and pliers become indispensable tools.
3. Larger wire sizes are problematic. The size makes the bend radius on smaller cores harder to achieve. When using no.12 or no. 14 wire, it is easier to wind on a core at least 2 in in diameter.

The wires of a ribbon are held in place with 3M no. 27 glass tape or no. 92 polyimide tape every 3/4 in. Sevick used glass tape in many applications, but 1/2 in wide polyimide tape works just as well. If you are going to buy only one (expensive) roll of tape, it makes sense to use polyimide material for all aspects of transformer construction. The ribbon is first placed on the outside diameter (perpendicular to it) of the toroid with about 1 1/2 in overhang. The larger part of the wire is bent downward 90°, and the ribbon is placed on the inner diameter and bent another 90°. Then the ribbon is placed back on the outer diameter. The large end is fed back (with some bending) through the inside of the core to complete the first turn. The thumb and a pair of pliers help in making the first turn tight to the core. After two or three turns, the winding is well anchored to the core and then can be completed quite easily.

Windings covered with PTFE tubing are even more problematic. The bend radius is harder to maintain, and the windings near each end tend to unwind. It is useful to use no. 92 tape to hold the beginning and ending of the windings to the core to maintain the tight winding.

13.5   Constructing Low Impedance Coax Cable

When designing transmission line transformers that require characteristic impedances less than 25 Ω, we usually choose rectangular line or low impedance coax cable. Neither of these transmission lines is readily available commercially in the range of 5–35 Ω. If one has access to a machine shop or a sheet metal shop, then a sheet of copper can be cut to the width required for the specific characteristic impedance. Characteristic impedances as low as 5 Ω have been obtained by Sevick using 3/8 in strips of copper and one layer of 3M no. 92 tape for insulation. A sheet metal shear is perfect for cutting copper sheet to the correct width. A sheet metal vendor is likely to have a shear available to cut your material to size for free or a nominal charge. With homemade, low impedance coax cable, values from 9 to 35 Ω have been easily constructed from readily available components. With further effort, values as low as 5 Ω should be achievable.

Low impedance coax cables have a decided advantage over tightly wound wire transmission lines because the currents can be considerably larger, since they are evenly distributed about the inner conductor and the outer braid; and because the voltage breakdowns are considerably larger (rivaling that of RG-8/U) since several layers of 3M no. 92 tape are generally used in achieving the desired characteristic impedances.

Table 4-1 shows the characteristic impedance as a function of various combinations of wire size and insulation thickness. Although not shown in the table, a single layer of 3M no. 92 tape on a no. 10 inner conductor was found to produce a characteristic impedance of 9 Ω. ML or H Imideze wire, without any extra insulation, should yield even lower impedances. The outer braids of all the cables in Table 4-1 were tightly wrapped with 3M no. 92 tape. Without this outer wrap, the characteristic impedance was found to be greater by about 25%.

An indispensable tool for constructing coax cables is shown in Figure 13-1. This U frame, mounted in a vise or clamped to a workbench, can be constructed out of any scrap wood on hand. Sevick used 1/4 in plywood. The jig shown was made from scraps of hobby wood and 1/2 in plywood already on hand from one of the local building supply stores and has a usable span from 15 to 30 in depending on how one attaches the vertical pieces using wood screws. The vertical struts are long enough that one can easily pass the roll of tape between the wire and the base. The horizontal

Figure 13-1 Photo shows a wooden jig used to stretch the inner conductor of handmade low impedance coax cable in preparation for wrapping the wire with polyimide tape.

piece is 24 in long. Basically, any type of attachment that will hold the wire taut will work. One end of the jig shown uses a hook and the other uses a no. 10 wood screw. The wire is wrapped around the hook and then hand stretched tight around the screw and held in place with two or three wraps of wire around the screw. Be sure to make the distance between the anchors at least 4 in longer than intended to allow for the lead length beyond the finished transformer windings. The insulation on the inner conductor can be put on by one of the following two methods:

1. *Longitudinal*: This is like rolling up a carpet or a window shade. 3M no. 92 or no. 27 is attached at one edge, along the length of the wire. The 1/2 in tapes, which are then rolled on, put about two layers on 12 to 16 gauge wire. If four layers are required, then this is done twice. As a result, the no. 16 wire will have a little more than four layers on it and the no. 12 wire a little less. Since the characteristic impedance is a function of the log of the ratio of the diameter of the outer braid to the diameter of the inner conductor (and therefore not sensitive to small differences in actual diameters), the desired result of two or four layers is practically realized. This is the method Sevick used.

2. *Spiral*: This is like taping a baseball bat. By carefully controlling the pitch of the spiral (and hence the overlap), it is possible to place, quite accurately, the desired number of layers. For four layers, it is recommended that this process be done in two steps. This method is also easy to use because of the tall struts

in the U frame. I have tried various tape widths for spiral application. The 1/2 in tape is by far the easiest to apply. One inch and larger sizes tend toward a simple 90° wrap after a few loops, which defeats the purpose of a spiral wrap.

After the insulation has been placed on the inner conductor, it is removed from the U frame and made ready for the outer braid. Except for the 30 to 35 Ω cables in Table 4-1, which use the outer braid from RG-58/U, all of the others use the braids from smaller cables such as RG-122/U. Flat 1/8 in braid, which is opened up easily using a pencil point, has also been used successfully. The only requirement on the outer braid is that it must have practically 100% coverage. That is, the inner conductor should not be seen when the cable is wound around a core. Small end caps of 1/8 in copper strip or 14 or 16 gauge wire provide adequate contacts to the braid. If the braid is to be tightly wrapped, the whole structure is placed back on the U frame and a spiral winding of 3M no. 92 tape (or practically any other tape) is applied. Tapping the inner conductor is made possible by using two sections of outer braids and soldering them together around the tap. (Even the center conductor of a coax cable can have a longitudinal potential gradient.)

13.6   The Care and Handling of Ferrite Transformers

Nickel-zinc ferrites (which are the ones to be used in power transmission line transformers) have bulk resistivity in the range of 10^5 to 10^9 Ω-cm. This means that they are excellent insulators and do not require extra insulation from an electrical standpoint. Even bare wire (as long as it doesn't touch its neighboring turns) can be used without any extra precautions. Coatings on the cores are mainly used for the mechanical protection of the wires. With automatic winding machines especially, the insulation can be harmed by rough surfaces and sharp edges, which introduces the possibility of a short circuit. With the smooth surfaces and well chamfered edges of modern toroids, the harm, during winding, is considerably reduced or even eliminated (especially when done by hand).

Another misconception regarding ferrite cores or beads is the need for extra protection from the environment. Since ferrite is a ceramic, there is practically no moisture adsorption on the surface or penetration of it below the surface. Even if there were, the properties of the ferrite would be unchanged. For years, Sevick used ferrite transformers mounted in mini-boxes with their covers in place without using self-tapping screws or tape along the exposed edges. These transformers have been subjected to all kinds of weather without any noticeable changes in performance. The only precaution is to keep the rain and snow off the transformers. The coiled transmission lines would not like this.

And finally, what happens to ferrites (which are brittle) when they are broken into several large pieces? If the pieces are large enough (and not too many) they can be glued together, and the core will perform as well as before. The precaution here is to glue the parts as tightly as possible in order to eliminate the high reluctance of a sizable gap between the ferrite pieces. Gelled cyanoacrylate glue is an excellent choice.

Appendix A
Reprint of Guanella Article

The text that follows is that from Guanella's original article in the Brown Bovieri Review. The figures are copies from the original. The article is reproduced with the enthusiastic support and permission of ABB (a successor to Brown Bovieri).

NEW METHOD OF IMPEDANCE MATCHING
IN RADIO-FREQUENCY CIRCUITS

Decimal Index 621.396.611.39

A new transformer method is described which is suitable both for matching circuits of unequal impedance and coupling symmetrical and unsymmetrical radio-frequency circuits. In contradistinction to conventional methods of impedance matching the frequency of the oscillations being transmitted can be varied over a wide range without the necessity of re-tuning.

The impedances of the individual circuits of radiofrequency equipment are frequently unequal. In order to obviate the reflections and losses involved by mismatching, special matching devices have to be inserted between such dissimilar circuits for the transmission of energy. For instance, matching is necessary between the tubes of a transmitter output stage with high load resistance and the low-impedance antenna transmission line or feeder system. In the case of low frequencies transformers with a corresponding turns ratio can be employed. By reason of the unavoidable leakage inductance of the coupled transformer coils, high frequencies generally involve tuning by means of additional condensers, and should the working frequency be varied, corresponding re-tuning is therefore entailed.

For impedance matching purposes a quarter-wave Lecher wire system having a surge impedance which is the geometric mean between the two impedances to be matched can likewise be employed. Such matching sections must naturally also be re-tuned in the event of the frequency being altered, to correspond to the changed wave-length. Small frequency deviations are, however, permissible when the impedance transformation takes place in several steps adjusted to the mean frequency. – Another method of matching, the line with exponential taper, permits large frequency variations without re-tuning, but has amongst other things the drawback of taking up a large amount of space.

Special couplers are also necessary for transition from symmetrical to unsymmetrical circuits, e.g. between the symmetrical output of a push-pull transmitter stage and a coaxial antenna cable with earthed sheathing. Here, too, variation of the frequency generally involves re-tuning.

A new coupler which obviates re-tuning is shown in Figure 1a. It comprises two superposed windings W_1 and W_2 separated by an insulating tube R. Given symmetrical currents i_1 (full-lined arrows) the magnetic fields produced by two closely-spaced superposed sections of conductor practically neutralize each other, i.e. the mutual inductance of two successive turns of a coil can be neglected, while it is possible to replace the two windings by two straight conductors having the same cross-section, length, and spacing as the two developed windings. This Lecher wire system is represented in the equivalent diagram (Figure 1b) by the equivalent line A.

On the other hand, with unsymmetrical currents i_2 (dotted arrows), the field vectors produced by two superposed sections of the conductors are added together, with the result that the mutual inductance between the individual turns of the coil becomes an important factor. The double-wire coil system behaves here like a conventional choke coil, represented in the equivalent diagram by B. In this diagram the symmetrical and unsymmetrical currents i_1 and i_2, respectively, are segregated by centre-tapped ideal transformers T. Given an adequate number of

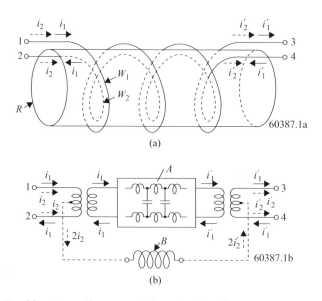

(a)

(b)

Figure 1 *Double-wire coil system with equivalent diagram.*
    (a) The coil system comprises two superposed windings W_1 and W_2
    separated by an insulating tube R.
    (b) According to this equivalent diagram. where symmetrical currents
    i_1 are concerned, the coil has the effect of a Lecher wire system A.
    but with unsymmetrical currents i_2 the nature of a choke coil B. The
    symmetrical and unsymmetrical currents are segregated by ideal
    centre-tapped transformers.

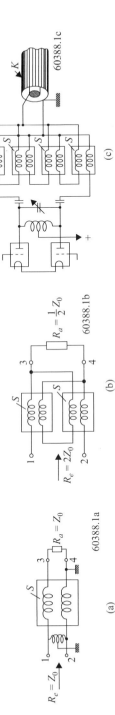

Figure 2   Employment of double-wire coil systems for coupling and Impedance matching purposes.

(a) Due to the suppression of the unsymmetrical currents by the series inductance of the coils such units can be used for coupling physically symmetrical circuits (connected to terminals 1 and 2) to circuits having one pole earthed (connected to terminals 3 and 4).

(b) By series-parallel connection of two coil systems S the load resistance $R_a = \frac{1}{2} Z_0$ Is transformed to the input impedance $R_e = 2Z_0$ ($Z_0 = $ surge impedance of a coil system).

(c) The antenna cable K and output stage are "matched" by the four coil systems S.
    Surge impedance of coil systems $= 240\ \Omega$.
    Surge impedance of cable $= 240\ \Omega : 4 = 60\ \Omega$.
    Load impedance of output stage $= 240\ \Omega \times 4 = 960\ \Omega$.

turns on the windings W_1 and W_2 the impedance of the equivalent choke coil B becomes so high that, even assuming unequal potentials between the centre tappings of the input and output coils, the unsymmetrical current i_2 can be neglected. In this case the described coil system forms an ideal transformer combined with an ideal line.

In view of the effect of this ideal transformer such a system S can now be employed, as shown for example in Figure 2a, to couple a physically symmetrical circuit (connected to terminals 1 and 2) to a load resistance R_a having one pole earthed. By making the coil of suitable dimensions the surge impedance Z_0 of the matching line *(A* in the equivalent diagram Figure 1b) represented by the coil system can be adapted to the pure load resistance R_a. In this case the input impedance R_e occurring between terminals 1 and 2 is equal to the surge impedance Z_0 and in consequence also to the load resistance R_a, immaterial of the actual working frequency.

By series-parallel connection of two or more coil systems impedance matching is now also possible in a simple manner, independent of the frequency. Figure 2b shows by way of example the input terminals of two systems of coils S connected in series and the output terminals in parallel. No objections can be raised to this

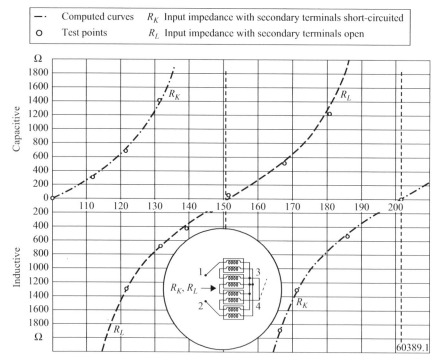

Figure 3   Input impedance of matching unit when output terminals short-circuited or open. The matching unit comprises four double wire coils in series-parallel connection. The computed and measured primary impedances are plotted as a function of the frequency with the secondary terminals open and short-circuited.

practice provided the series inductance (B in the equivalent diagram in Figure 1b) is large enough. The load resistance $R_a = \frac{1}{2}Z_0$ is thus transformed to the input impedance $2Z_0$. Analogously, with n coil systems impedance transformation in the ratio $1:n^2$ can be achieved.

In Figure 2c, for instance, four coil systems are shown connected between a transmitter output stage and the high-frequency antenna cable K, the resulting impedance transformation being in the ratio $4^2:1 = 16:1$. With a coil system having a surge impedance $Z_0 = 240\ \Omega$, for example, a transmitter output stage with a load impedance of $4 \times Z_0 = 960\ \Omega$ can be coupled to an antenna cable of $Z_0:4 = 60\ \Omega$. The coupled coil systems have the same effect as a transformer with separate windings, i.e. the symmetry of the anode circuit at the input end is not affected by single-pole earthing of the cable connected to the other end. Furthermore, the coupled coil systems behave like a Lecher wire system, i.e. the input impedance must follow a tangential function of the frequency when the terminals at the other end are open or short-circuited.

The curves in Figure 3 give the input impedance computed from the coil dimensions for conditions of short circuit and no-load. The measured impedance values are also given and agree with the curves to a high degree. These measurements, which demand great care, were made by a method specially developed for the purpose. The characteristic surge impedance can be determined from the geometric mean of the measured or computed short-circuit and no-load input impedances. In the present case it is about $240\ \Omega$. Figure 4 gives the curve of the input

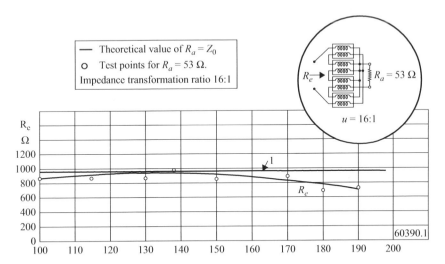

Figure 4   *Theoretical and measured Input Impedance of a matching unit with a pure resistive load. The matching unit comprises four double-wire coils in series-parallel connection. A surge Impedance of 240 Ω was computed from the coil data and the measurements in Figure 3, whence, assuming a pure resistive load of 60 Ω, the theoretical value of the input impedance is 960 Ω. The measured values of the input Impedance are somewhat lower owing to the load impedance having been somewhat lower than theoretically assumed.*

Figure 5   Matching unit with double-wire coils. The system contains four double-wire coils for impedance transformation from 60 Ω to about 1000 Ω in the case of metre waves. With a power of over 100 W the losses are negligible.

impedance for a load impedance of about 53 Ω. From the test points it is clear that the desired impedance transformation in the ratio of 1:16 is actually possible over a very wide frequency range. The deviation of the plotted mean-value curve R_e from the theoretical curve 1 is due to the load impedance being slightly lower than the theoretical value, as well as to the inherent capacitance of the circuit.

The described method of matching is particularly suitable for application in the ultra-short-wave field, where it represents a big simplification compared to conventional tuned matching devices. Figure 5 shows the external appearance of an impedance transformer with four coils, employed as antenna coupler in a medium-power transmitter. It requires little space and its losses are very low. This new component greatly simplifies the construction and operation of the equipment marketed by the Company.

(MS 564)   *G. Guanella. (E.G. W.)*

Appendix B
Some Broad-Band Transformers[*]
C. L. Ruthroff[†] Member, IRE

Summary—Several transmission line transformers are described which have band-width ratios as high as 20,000:1 in the frequency range of a few tens of kilocycles to over a thousand megacycles. Experimental data are presented on both transformers and hybrid circuits.

Typical applications are: interstage transformers for broad-band amplifiers; baluns for driving balanced antennas and broad-band oscilloscopes; and hybrids for use in pulse reflectometers, balanced modulators, etc.

These transformers can be made quite small. Excellent transformers have been made using ferrite toroids having an outside diameter of 0.080 inch.

Several transmission line transformers having bandwidths of several hundred megacycles are described here. The transformers are shown in Figures 1–9. When drawn in the transmission line form, the transforming properties are sometimes difficult to see. For this reason, a more conventional form is shown with the transmission line form. Some winding arrangements are also shown. Certain of these configurations have been discussed elsewhere and are included here for the sake of completeness [1–4].

In conventional transformers the interwinding capacity resonates with the leakage inductance producing a loss peak. This mechanism limits the high frequency response. In transmission line transformers, the coils are so arranged that the interwinding capacity is a component of the characteristic impedance of the line, and as such forms no resonances which seriously limit the bandwidth. Also, for this reason, the windings can be spaced closely together maintaining good coupling. The net result is that transformers can be built this way which have good high frequency response. In all of the transformers for which experimental data are presented, the transmission lines take the form of twisted pairs. In some configurations the high frequency response is determined by the length of the windings and while any type of transmission line can be used in principle, it is quite convenient to make very small windings with twisted pairs.

*Original manuscript received by the IRE, February 5, 1959; revised manuscript received, April 1, 1959.
†Bell Telephone Labs., Inc., Holmdel, N. J.

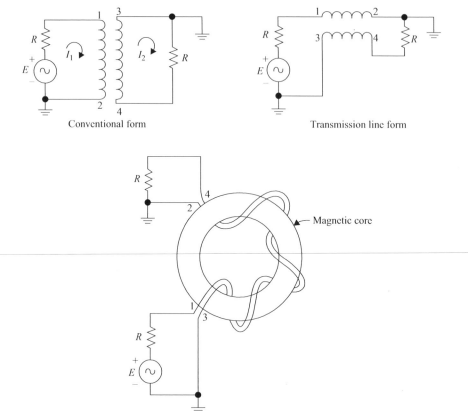

Figure 1   Reversing transformer.

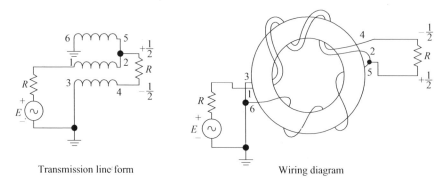

Transmission line form   Wiring diagram

Figure 2   Unbalanced to balanced transformer.

The sketches showing the conventional form of transformer demonstrate clearly that the low frequency response is determined in the usual way, i.e., by the primary inductance. The larger the core permeability, the fewer the turns required for a given low frequency response and the larger the over-all bandwidth. Thus a

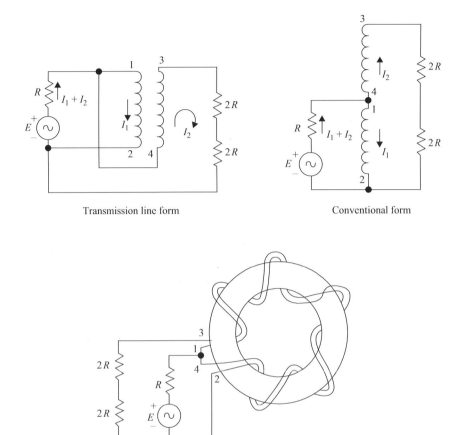

Figure 3   4:1 Impedance transformer.

good core material is desirable. Ferrite toroids have been found very satisfactory. The permeability of some ferrites is very high at low frequencies and falls off at higher frequencies. Thus, at low frequencies, large reactance can be obtained with few turns. When the permeability falls off the reactance is maintained by the increase in frequency and good response is obtained over a large frequency range. It is important that the coupling be high at all frequencies or the transformer action fails. Fortunately, the bifilar winding tends to give good coupling. All of the cores used in the experimental transformers described here were supplied by F. J. Schnettler of the Bell Telephone Laboratories, Inc.

Polarity Reversing Transformer-Figure 1

This transformer consists of a single bifilar winding and is the basic building block for all of the transformers. That a reversal is obtained is seen from the conventional

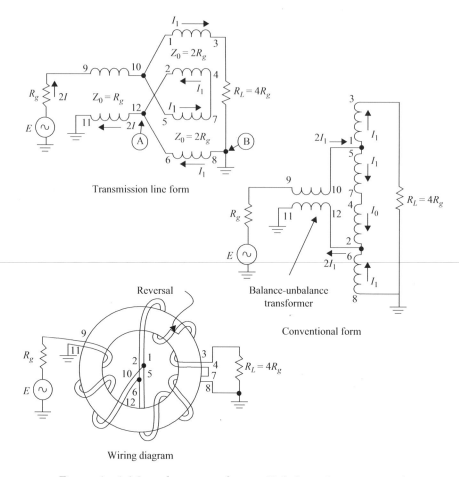

Transmission line form

Balance-unbalance
transformer

Conventional form

Reversal

Wiring diagram

Figure 4   4:1 Impedance transformer. Unbalanced—symmetrical.

form which indicates current polarities. Both ends of the load resistor are isolated from ground by coil reactance. Either end of the load resistor can then be grounded, depending upon the output polarity desired. If the center of the resistor is grounded, the output is balanced. A suitable winding consists of a twisted pair of Formex insulated wire. In such a winding, the primary and secondary are very close together, insuring good coupling. The interwinding capacity is absorbed in the characteristic impedance of the line.

At high frequencies this transformer can be regarded as an ideal reversing transformer plus a length of transmission line. If the characteristic impedance of the line is equal to the terminating impedances, the transmission is inherently broadband. If not, there will be a dip in the response at the frequency at which the transmission line is a quarter-wavelength long. The depth of the dip is a function of the ratio of terminating impedance to line impedance and is easily calculated. Experimental data

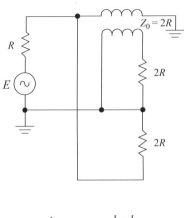

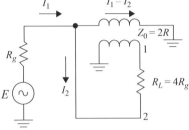

Figure 5   Balanced—unbalanced 4:1 impedance transformer.

on a reversing transformer are shown in Figures 10 and 11. Figure 10 is the response of a transformer with no extra impedance matching. The return loss of this transformer to a 3 mμsec pulse is 20 db. The transformer of Figure 11 has been adjusted to provide more than 40 db return loss to a 3 mμsec pulse. The transformer loss (about 0.5 db before matching) is matched to 75 ohms with the two 3.8-ohm resistors. The inductance is tuned out with the capacity of the resistors to the ground plane. The match was adjusted while watching the reflection of a 3 mμsec pulse.

Balanced-To-Unbalanced 1:1 Impedance Transformer-Figure 2

This is similar to Figure 1 except that an extra length of winding is added. This is necessary to complete the path for the magnetizing current.

Unbalanced-Unsymmetrical 4:1 Impedance Transformer-Figure 3

This transformer is interesting because with it a 4:1 impedance transformation is obtained with a single bifilar winding such as used in the reversing transformer.

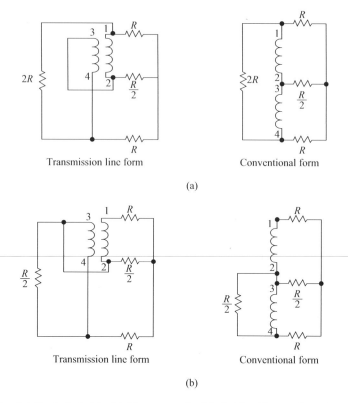

*Figure 6   (a) Basic hybrid. (b) Unsymmetrical hybrid with equal conjugate
impedances.*

The transforming properties are evident from Figure 3. Not so easily seen is the
high frequency cutoff characteristic.

The response of this device at high frequencies is derived in the Appendix and
only the result for matched impedances is given here.

$$\frac{\text{Power Available}}{\text{Power Output}} = \frac{(1 + 3\cos\beta l)^2 + 4\sin^2\beta l}{4(1 + \cos\beta l)^2} \tag{1}$$

where β is the phase constant of the line, and l is the length of the line. Thus, the
response is down 1 db when the line length is $\lambda/4$ wavelengths and the response is
zero at $\lambda/2$. For wideband response this transformer must be made small. For a plot
of (1) see Figure 16.

Experimental data are given for a transformer of this type in Figure 12.

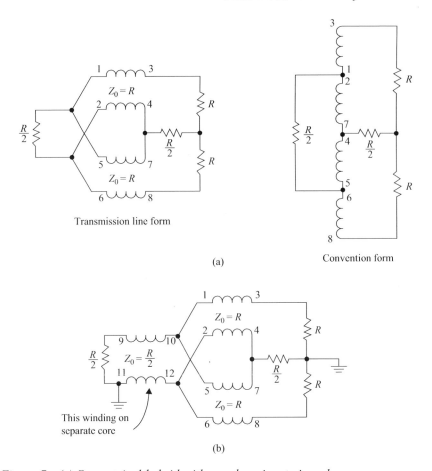

Figure 7 (a) Symmetrical hybrid with equal conjugate impedances.
(b) Unbalanced symmetrical hybrid with equal conjugate impedances.

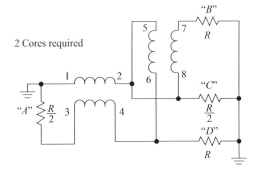

Figure 8 Hybrid with equal conjugate impedances. Each arm single ended.

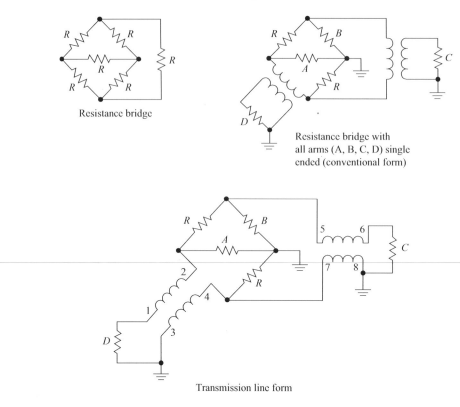

Figure 9   Resistance hybrid with equal impedance loads. (This hybrid has 3 db loss in addition to transformer loss.)

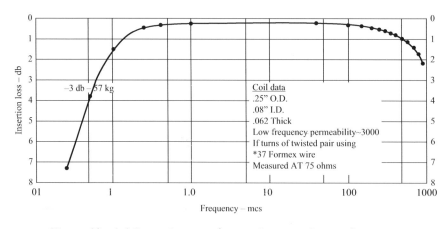

Figure 10   1:1 Reversing transformer. Insertion loss vs frequency.

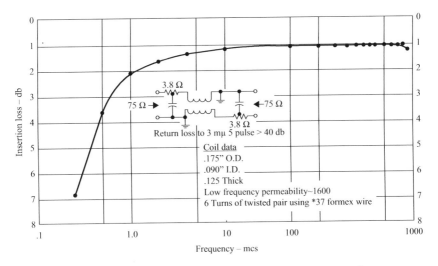

Figure 11   Matched reversing transformer. Insertion loss vs frequency.

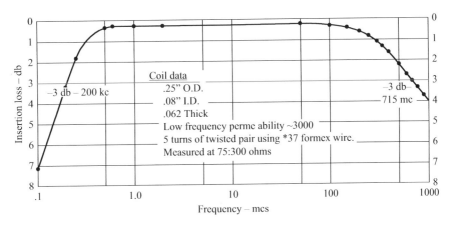

Figure 12   Winding 4:1 impedance transformer unbalanced—unsymmetrical.

Unbalanced-Symmetrical 4:1 Impedance Transformer—Figure 4

This configuration requires three bifilar windings as shown in Figure 4. All three windings can be placed on one core, a procedure which improves the low frequency response.[1] When winding multiwinding transformers the following well-known rule should be followed: with the generator connected and the load open, a completed circuit should be formed by the windings so that the core will be magnetized. The fields set up by the currents should be arranged so as to aid each other.

[1]Pointed out to the author by N. J. Pierce of Bell Telephone Labs., Inc., Holmdel, N. J.

Balanced-To-Unbalanced 4:1 Impedance Transformers–Figure 5

The circuit of Figure 5 is quite simple. The single bifilar winding is used as a reversing transformer as in Figure 1. The high frequency cutoff is the same as that for the transformer of Figure 3.

In some applications it is desirable to omit the physical ground on the balanced end. In such cases, Figure 5(b) can be used. The high frequency cutoff is the same as for the transformer of Figure 3. The low frequency analysis is presented in Appendix B.

Hybrid Circuits: Figures 6-9

Various hybrid circuits are developed from the basic form using the transformers discussed previously. The drawings are very nearly self-explanatory. In all hybrids in which all four arms are single-ended, it has been found necessary to use two cores in order to get proper magnetizing currents.

Two hybrids have been measured and data included here. The response of a hybrid of the type shown in Figure 8 is given in Figure 13. For this measurement $R = 150$ ohms. In order to measure the hybrid in a 75-ohm circuit, arms B, D were measured with 75-ohm series resistances in series with the 75-ohm measuring gear. This accounts for 3 db of the loss. Under these conditions arms B and D have a 6 db return loss.

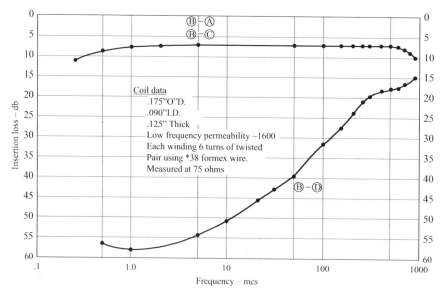

Figure 13   Hybrid of Figure 8. Insertion loss vs frequency.

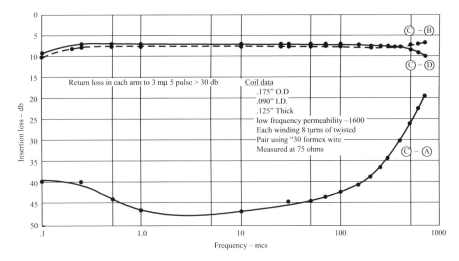

Figure 14   Matched resistance hybrid. Insertion loss vs frequency.

The transmission of the resistance hybrid of Figure 9 is given in Figure 14. This hybrid has been matched using the technique described previously for the reversing transformer. The results of this matching are included in the figure. This hybrid was designed for use in a pulse reflectometer, the main part of which is a stroboscopic oscilloscope with a resolution of better than 3 mμsec. The oscilloscope was designed by W. M. Goodall.

Applications

Many applications for these transformers will occur to the reader. For purposes of illustration, a few of them are listed here.

1. The reversing transformer of Figure 1 can be used to reverse the polarity of short pulses, an operation which is frequently necessary. It has also been used in balanced detectors and to drive push-pull amplifiers from single-ended generators.
2. The transformers of Figures 2 and 5(b) are useful for driving balanced antennas. The circuit of Figure 5(b) may find application in connecting twin lead transmission line to commercial television receivers.
3. The transformer of Figure 3 has found wide use in broadband amplifier interstages. It will also be useful in transforming the high output impedances of distributed amplifiers to coaxial cable impedances. They can also be cascaded to get higher turns ratios.
4. The circuit of Figure 5(a) has been used to drive broadband oscilloscopes, with balanced inputs, from single-ended generators. It can also find use in balanced detectors.

5. Hybrids have many uses such as in-power dividers, balanced amplitude and phase detectors; as directional couplers for pulse reflectometers, IF and broadband sweepers. They might also be used as necessary components in a short pulse repeater for passing pulses in both directions on a single transmission line.

Appendix A

The high frequency response of the circuit of Figure 3 is derived from Figure 15. The loop equations are as follows:

$$e = (I_1 + I_2) R_g + V_1$$
$$e = (I_1 + I_2) R_g - V_2 + I_2 R_L$$
$$V_1 = V_2 \cos \beta l + j\, I_2 Z_0 \sin \beta l \tag{2}$$
$$I_1 = I_2 \cos \beta l + j\, \frac{V_2}{Z_0} \sin \beta l.$$

This set of equations is solved for the output power P_0.

$$P_0 = |I_2|^2 R_L = \frac{e^2 (1 + \cos \beta l)^2 R_L}{\left[+2R_g(1 + \cos \beta l) \right]^2 + \left[\frac{R_g R_L + Z_0^2}{Z_0} \right]^2 \sin^2 \beta l} . \tag{3}$$

From this expression, the conditions for maximum power transmission are obtained by setting $l = 0$ and setting $dP_0/dR_L \big|_{l=0} = 0$. The transformer is matched when $R_L = 4R_g$. The optimum value for Z_0 is obtained by minimizing the coefficient of $\sin^2 \beta l$ in (3). In this manner the proper value for Z_0 is found to be $Z_0 = 2R_g$.

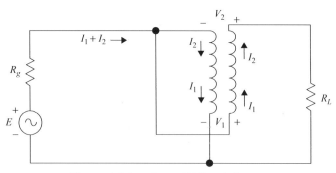

Characteristic impedance of bifilar winding $= Z_0$
the reactance of the windings $X \gg R_L, R_g$

Figure 15　Transformer schematic.

Now, setting $R_L = 4R_g$ and $Z_0 = 2R_g$, (3) reduces to

$$P_0 = \frac{e^2(1 + \cos \beta l)^2}{R_g\left[(1 + 3 \cos \beta l)^2 + 4 \sin^2 \beta l\right]} . \tag{4}$$

Also,

$$P_{\text{available}} = \frac{e^2}{4R_g}, \tag{5}$$

and dividing (4) by (3):

$$\frac{\text{Power Available}}{\text{Power Output}} = \frac{(1 + 3 \cos \beta l)^2 + 4 \sin^2 \beta l}{4(1 + \cos \beta l)^2} \tag{6}$$

This function is plotted in Figure 16.

The impedances seen at either end of the transformer with the other end terminated in Z_L have been derived. They are:

$$Z_{\text{in}} \text{ (low impedance end)} = Z_0\left(\frac{Z_L \cos \beta l + jZ_0 \sin \beta l}{2 Z_0(1 + \cos \beta l) + j Z_L \sin \beta l}\right) \tag{7}$$

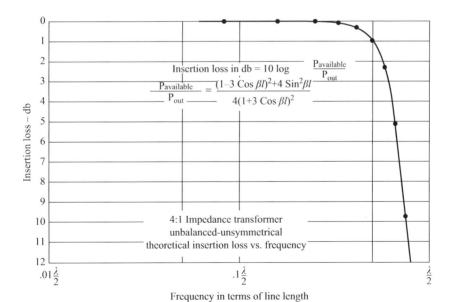

Figure 16   Theoretical insertion loss vs frequency.

and

$$Z_{in} \text{ (high impedance end)} = Z_0 \left(\frac{2Z_L(1 + \cos \beta l) + jZ_0 \sin \beta l}{Z_0 \cos \beta l + j Z_L \sin \beta l} \right) \tag{8}$$

Appendix B

In the low frequency analysis of the transformer of Figure 5 the series impedance of each half of the bifilar winding is denoted by Z. The loop equations are:

$$E = (R_g + Z)I_1 - (Z + kZ)I_2$$
$$E = (R_g - kZ)I_1 + (R_L + Z + kZ)I_2 \tag{9}$$

from which

$$\frac{I_1}{I_2} = \frac{R_L + 2Z(1 + k)}{Z(1 + k)} \approx 2 \quad \text{if } Z \gg R_L \tag{10}$$

We now proceed to calculate the voltages from points 1 and 2 to ground

$$V_{2G} = E = I_1 R_g$$

When the transformer is matched, $E = 2 I_1 R_g$ and

$$V_{2G} = I_1 R_g \tag{11}$$

Similarly,

$$V_{1G} = I_2 Z - kZ(I_1 - I_2).$$

With the aid of (10) this can be rearranged to

$$V_{1G} = Z I_1 \left[\frac{Z(1 + k)^2 - kR_L - 2kZ(1 + k)}{R_L + 2Z(1 + k)} \right]. \tag{12}$$

Now let the coupling coefficient $k = 1$, then

$$V_{1G} = I_1 Z \left[\frac{-kR_L}{R_L + 2Z(1 + k)} \right] \approx -\frac{I_1 R_L}{4} \quad \text{for } Z \gg R_L$$

When the transformer is matched, $R_L = 4R_g$ so that

$$V_{1G} = I_1 R_g = -V_{2G}, \tag{13}$$

and the load is balanced with respect to ground.

From (13) it is clear that the center point of R_L is at ground potential. This point can therefore be grounded physically, resulting in Figure 5(a).

Acknowledgement

In addition to those mentioned in the text, the author is indebted to D. H. Ring for many stimulating discussions on every aspect of these transformers.

References

[1]   Wollmor K. Roberts, "A new wide-band balun," Proc. IRE, vol. 45, pp. 1628-1631; December 1957.

[2]   H. Gunther Rudenberg, "The distributed transformer," Raytheon Mfg. Co. Waltham, Mass.

[3]   G. Guanella, "New method of impedance matching in radio frequency circuits," *Brown-Bovieri Rev.*, vol. 31, p.327; 1944.

[4]   A. I. Talkin and J. V. Cuneo, "Wide-band balun transformer," *Review of Sci. Inst.*, vol. 28, No. 10, pp. 808-815; October, 1957.

[5]   C. A. Burrus, unpublished memorandum.

Note that the original manuscript does not have a reference [5] in the text of the article.

Index